Semantic Modeling for the Acquisition of Topographic Information from Images and Maps

SMATI 97

Edited by
Wolfgang Förstner
Lutz Plümer

Birkhäuser Verlag
Basel · Boston · Berlin

Editors:

Prof. Dr.-Ing.
Wolfgang Förstner
Institut für Photogrammetrie
Universität Bonn
Nussalle15
D-53115 Bonn

Priv.-Doz. Dr.
Lutz Plümer
Institut für Informatik III
Universität Bonn
Römerstr. 164
D-53117 Bonn

Deutsche Bibliothek Cataloging-in-Publication Data

Semantic modeling for the acquisition of topographic information from images and maps / SMATI 97. Ed. by Wolfgang Förstner ; Lutz Plümer. – Basel ; Boston ; Berlin : Birkhäuser, 1997
ISBN 3-7643-5758-4

Camera-ready Vorlage durch die Autoren erstellt
Gedruckt auf säurefreiem Papier, hergestellt aus chlorfrei gebleichtem Zellstoff
Signet: Ingo Petzold / Monika Sester
Printed in Germany
ISBN 3-7643-5758-4

9 8 7 6 5 4 3 2 1

Table of Contents

Buildings

Preface

Acquiring spatial data for Geoinformationsystems still is mainly done by human operators who analyse images using classical photogrammetric equipment or digitise maps, possibly assisted by some low level image processing. Automation of these tasks is difficult due to the complexity of the object, the topography, and the deficiency of current pattern recognition and image analysis tools for achieving a reliable transition from the data to the high level description of topographic objects. It appears that progress in automation only can be achieved by incorporating domain specific semantic models into the analysis procedures.

This volume collects papers which were presented at the Workshop "SMATI '97". The workshop focused on the "Semantic Modelling for the Acquisition of Topographic Information from Images and Maps". It was held at the Gustav Stresemann Institute in Bonn Bad Godesberg, May 21 - 23, 1997. The workshop brought together specialists in Photogrammetry, Cartography, Computer Vision, and Image Understanding in order to discuss the current state of automatic procedures for acquiring topographic information from images and maps. The goal was to find out common object models and analysis strategies.

The idea for this Workshop originated in the joint research project "Semantic Modelling". This initiative, being funded by the German Research Council since 1994, brings together researchers from Photogrammetry, Cartography, Image Understanding and Pattern Recognition. Its common aim is the modelling of object acquisition, description and use in Geoinformation. The concept *Semantic Modelling* is used as link for integrating the techniques to be developed. Different scientific disciplines are integrated through the development of joint models for topographic objects, the development of a joint conceptual and methodological framework for the development of the tools, and a close co-operation with respect to both methodology and application.

The papers collected in this proceedings were carefully reviewed and selected in a doubly blind process by an international program committee consisting of

Mike Brady, Oxford
Leonie Dreschler-Fischer, Hamburg
Andrew Frank, Wien
Armin Grün, Zürich
Henri Maitre, ENST
Chris McGlone, CMU
Bernd Neumann, Hamburg
Ram Nevatia, USC
Heinrich Niemann, Erlangen
Ed Riseman, UMASS
Gerhard Sagerer, Bielefeld
Robert Weibel, Zürich
George Vosselman, Delft

We have structured the contents of this volume into four themes, namely Methods, Cartography, Roads and Buildings which have also been the key issues of the workshop.

We would like to thank all individuals who contributed to the preparation of this volume and the organisation of the workshop, namely the members of the program committee, and the authors of the presented papers.

Wolfgang Förstner
Institut für Photogrammetrie
Universität Bonn

Lutz Plümer
Institut für Informatik III
Universität Bonn

Methods

Semanitic Modeling for the Acquisition of Topografic
Information from Images and Maps
ed. by W. Förstner and L. Plümer

Knowledge Based Interpretation of Aerial Images and Maps Using a Digital Landscape Model as Partial Interpretation

Hans Koch[1], Kian Pakzad[2], Ralf Tönjes[3]

[1]Institut für Kartographie
Universität Hannover, Appelstr. 9A, 30167 Hannover, Germany
Ph.: +49-511-762-4967, Fax +49-511-762-2780,
e-mail: koch@jakob.ifk.uni-hannover.de

[2]Institut für Photogrammetrie und Ingenieurvermessungen
Universität Hannover, Nienburger Straße 1, 30167 Hannover, Germany
PH.: +49-511-762-3893, Fax: +49-511-762-2483
e-mail: pakzad@ipi.uni-hannover.de

[3]Institut für Theoretische Nachrichtentechnik und Informationsverarbeitung
Universität Hannover, Appelstr. 9A, 30167 Hannover, Germany
PH.: +49-511-762-5326, Fax: +49-511-762-5333
e-mail: toenjes@tnt.uni-hannover.de

Abstract

The methods for the interpretation of aerial images and maps are usually different although both describe the same landscape. The presented work shows that regarding remote sensing data and maps as different kinds of sensors allows a similar approach for both in the domain of landscape interpretation. The prior knowledge about the landscape objects is represented explicitly by semantic nets. Based on the semantics of the network language a problem independent set of rules controls the scene interpretation.

Important is that the scene analysis employs a partial interpretation derived from a Digital Landscape Model (ATKIS DLM 25/1). This partial interpretation is used to generate an initial scene description. Consecutively the scene description is verified in aerial images and maps. Interpretation proceeds iteratively mixing top-down and bottom-up strategies.

This paper shows the representation of the knowledge in several semantic layers, the strategies of interpretation, and the methods to interprete aerial images and maps using a DLM as partial interpretation.

1 Introduction

The recognition of land use changes for map updating and environmental and agricultural monitoring represents a major topic of remote sensing. Long term changes of the environment can be obtained from the topographic map 1:25000 (TK 25) that is updated since about one hundred years. Due to the large amount of data photogrammetry and cartography work on the automatic extraction of objects from sensor data and maps. However, until now no prior knowledge about already mapped landscapes is utilized.

This paper presents an approach that integrates an existing partial interpretation derived from a GIS (ATKIS DLM 25/1) in the recognition process. The landscape is modelled in a semantic net, which contains the representation of the landscape in ATKIS and in different aerial images and maps. Moreover, the model contains query operations for ATKIS data and methods for extracting primitives from aerial images and maps.

In the first interpretation step an initial scene description is instantiated according to the ATKIS data. Consecutively the derived scene description is verified in aerial images and maps and adapted if necessary.

2 System Overview

Figure 1 shows the architecture of the knowledge based scene analysis system. The system is designed for the interpretation of aerial images and maps.

The prior knowledge about the objects to be extracted from the image data is represented explicitly in the knowledge base. Besides this general knowledge about the objects the interpretation takes advantage of a digital landscape model (ATKIS DLM 25/1) that contains the object location for some object classes. From the DLM a partial interpretation of the scene is derived. However the DLM has a limited accuracy and may be out-dated. Hence it has to be verified in the aerial images and maps that mirror the scene at the date of interest.

The interpretation module establishes an initial symbolic scene description based on the DLM which constitutes a hypothesis. This hypothesis is tested in the sensor data by

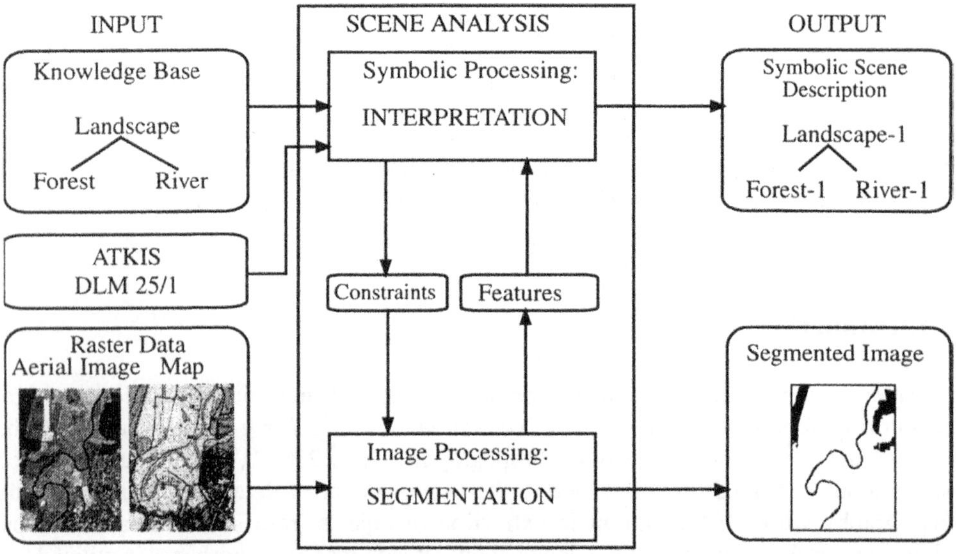

Fig. 1: Architecture of the knowledge based scene analysis system

generating constraints for the features expected in the image.

The image processing module extracts features that meet the constraints given by the interpretation module. It returns the found features and a reliability measure. The interpretation module groups the features and evaluates the relationship between the features to decrease or increase the reliability of competing interpretations. The interpretation results in a symbolic scene description and a corresponding segmented image.

3 A Language for Image Interpretation

3.1 Knowledge Representation

In the literature various systems for knowledge based scene analysis have been suggested. They employ different paradigms for knowledge representation like formal logic, fuzzy logic, frames (Clement et al., 1993; Foresti et al.; 1993), semantic nets (Niemann et al., 1990), and production systems (Matsuyama and Hwang, 1990; McKeown et al., 1985). Often knowledge about the objects and the control strategy is structured hierarchically (Foresti et al., 1993; Matsuyama and Hwang, 1990).

To cope with imprecision and uncertainty of sensor data the analysis systems has to combine bottom-up and top-down image analysis strategies, like SIGMA (Matsuyama and Hwang, 1990) and ERNEST (Niemann et al., 1990).

Usually the analysis results in many competing hypotheses. To reduce the search space analysis has to consider only the most promising hypotheses. To solve this indexing problem search methods are employed, like the A*-algorithm (Niemann et al., 1990) or production rules (Matsuyama and Hwang, 1990; McKeown et al., 1985) which exploit heuristics.

Most image analysis systems have been tested for selected examples, like the recognition of airports (McKeown et al., 1985), roads (Clement et al., 1993; Matsuyama and Hwang, 1990; Mayer and Steger, 1996), and buildings (Matsuyama and Hwang, 1990; Braun et al., 1995) in aerial images. A good overview about knowledge based scene analysis systems can be found in (Shapiro, 1992) and (Haralick and Shapiro, 1993). However, the data sets were small. Furthermore, image analysis should be able to take advantage of existing partial interpretations, i.e. GIS-data.

To ease the adaptation of the knowledge about the objects as well as the analysis strategy for new modelling tasks, the knowledge base has to be formulated explicitly.

Structural knowledge, like the knowledge about the relationships between the objects and their connection to the features apparent in the image data, can be represented efficiently by semantic nets. This kind of knowledge is termed fact knowledge. The procedural knowledge dictates how the fact knowledge will be used to control image interpretation.

Fact Knowledge is represented by semantic nets. Semantic nets consist of nodes and edges between them. Here, the nodes of the semantic net represent the objects in the scene or their sensor specific features respectively. The nodes are implemented as frames which contain a collection of attributes. Furthermore, an object has methods, i.e. functions, assigned to it to compute the attribute values. Additionally, an object can possess a method to segment the object in the image data.

Here the syntax of the semantic net distinguishes between two types of nodes: concepts and instances. The concepts describe the generic model of the objects. The instances are realizations of the concept in the observed scene. During interpretation four different states of the object recognition are distinguished: hypotheses I_H, missing instances I_M, partial instances I_P and complete instances I_C. Hypotheses are not yet verified in the sensor data. Falsification in the sensor data results in missing instances, while verification obtains partial and complete instances successively. Complete instances possess all obligatory parts of the object. Partial instances describe a predecessor state that contains already all obligatory parts that are not context dependent.

Interpretation starts with hypotheses instances which are initialized with the attributes of the concept. Methods assigned to the object attributes restrict the expected range of attributes top-down and obtain the measured value bottom-up from the sensor. To model uncertainties the attributes are described by minimum and maximum values and ranges. The differences between the expected range and the measured value is judged by a compatibility measure.

The nodes are related by different types of edges or links. The instances are related via *instance-of* to their concepts.

The specialization of an object is described by the *is-a* link. This link type introduces the concept of inheritance.

Objects are composed of parts, indicated by the *part-of* link. Object search can be reduced to a more simple task, the detection of its components. Furthermore some parts may only be detectable if other parts have already been detected and thus established a certain context. These context dependent parts are modelled by *cdpart-of*.

Objects can often be detected based on their geometric or photometric appearance, that can directly be segmented in the image data. This transformation of an abstract symbol to a concrete realization is represented by the concrete-of link, abbreviated *con-of*. The *con-of* link allows to distinguish between different conceptual layers, e.g. geometry and sensor.

The *data-of* link establishes a relation to the features segmented in the image data. A present *data-of* link indicates that the object can be segmented directly in the sensor data by image processing operators.

Geometric or photometric relations between objects can be represented explicitly by *attributed relations*. This link type contains an attribute that restricts the attribute value of the related object. While the *is-a*, *part-of*, and *con-of* relations propagate information top-down or bottom-up the *attributed relation* propagates information mainly horizontally.

Counters in the links of the concept net state how many links are required, i.e. obligatory, and allowed, i.e. optional, in the instance net.

Procedural Knowledge states how and in which order scene analysis has to proceed. A strategy is represented by a set of problem independent rules. These rules exploit only the semantics of the network language. The advantage of this approach is that the rules are independent from the domain knowledge. According to the number of link types only a small set of rules is required.

A rule is composed of a condition and an action part. The condition checks for a new interpretation state of neighboured nodes in the semantic net. The action part adapts the interpretation state of the focused node accordingly. The knowledge, for example, that an object n_0 is detected and can be denoted as complete instance I_C, if all its parts $n_i \in P$ are detected, is represented by following rule:

Rule-complete-instantiation:

CONDITION: If state (node n_i) = complete instance

$\forall n_i \in P \; P = \{n_i | n_i = \text{part-of}(n_0)\}$

ACTION: Then state(node n_0) = complete instance.

The sequence of the instantiation process is controlled by the priority of the rules. Rules which verify a node as partial or complete instance or falsify as missing instance have a higher priority than rules which generate new hypotheses. To favour top-down control the rules for top-down propagation of hypotheses are tested before bottom-up hypothesis propagation. Furthermore obligatory parts are searched for first followed by context dependent and optional parts. Attributed relations constrain the search space. Hence the associated rules have a high priority.

The rule based formulation of the control strategy eases the development and testing of new control strategies.

3.2 Control of Interpretation

The aim of image interpretation is to assign a symbolic description, i.e. semantics, to the data. Image interpretation exploits the generic model to instantiate hypotheses of objects expected in the scene. Interpretation is complete when an instance of the goal concept is found based on the generic model net and the features, which are segmented in the image data.

Figure 2 shows the interpretation process. The interpretation exploits the knowledge base to derive a symbolic scene description from the input data. Each possible interpretation is documented by a search node which contains all concepts and instances with their current interpretation state. If competing interpretations occur the search node splits into child search nodes. The leaves of the resulting search tree represent the currently competing interpretations. To focus interpretation on promising search nodes they are judged and ranked. The judgement computes the compatibility between expected properties and

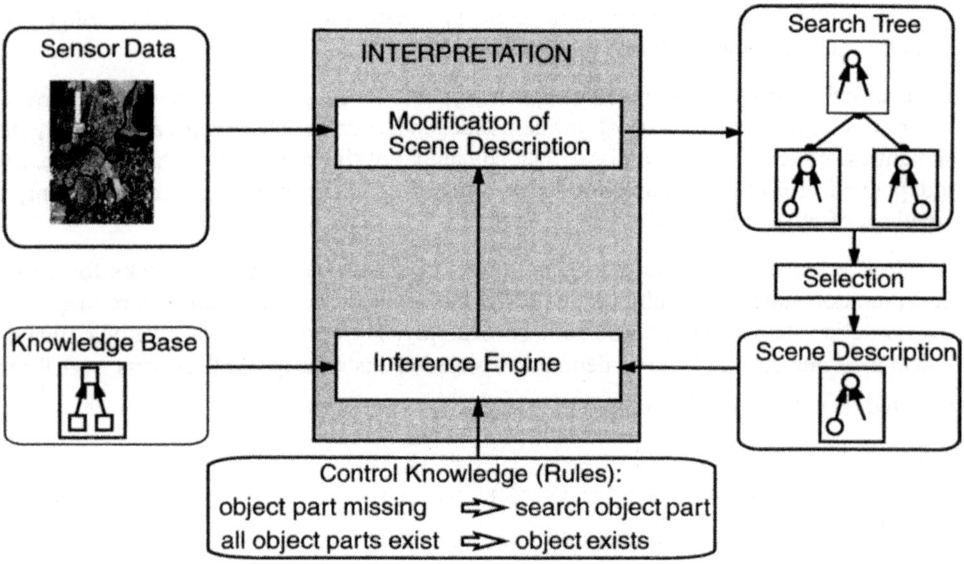

Fig. 2: Interpretation process

found properties of the scene description. An A*-algorithm selects the interpretation judged best for further investigation.

The instance net of the selected scene description is compared with the concept net in the knowledge base. The condition part of the rules takes care of the comparison. Interpretation employs an inference engine which controls the execution of the rules. According to the set of rules and their priority different strategies are possible. Usually the comparison matches for the condition of more than one rule. To select one rule the rules are ranked. The rank of a rule depends primarily on the priority of the rule in the selected strategy and secondly on the rank of the node for which the rule matches. The rank of the node corresponds to the distance of the node to the bottom data nodes, measured in the number of required hops via the links. Here, to favour immediate verification in the sensor data the nodes are ranked bottom-up.

The selected rule is executed and modifies the scene description by establishing new links or changing the status of a node. Competing modifications of the scene description cause the current search node to split for each possible interpretation. Competing interpretations occur if

- a segmentation method returns more than one feature that meets the request,
- a node is specialized, or
- a node establishes links to other nodes that are exclusive.

Here, to avoid an explosion of the search tree, hypotheses are primarily propagated top-down from the scene layer to the sensor layer to verify them in the sensor data. At the sensor layer the propagated hypothesis calls methods for segmentation. The result of the verification is returned to the superior instance which consecutively generates new hypotheses.

3.3 Implementation

The described approach is implemented as a knowledge based interpretation system called AIDA. The control algorithms of the system are implemented in C++ with a Tcl/Tk interpreter interface. A graphical user interface provides browser and editor functionality for monitoring and knowledge acquisition.

4 Using GIS Data as Partial Interpretation

4.1 ATKIS

The partial interpretation is derived from the 'Authoritative Topographic Cartographic Information System (ATKIS)' (Grünreich, 1992), which has been developed by the German state Ordnance Surveys. ATKIS data is available in most states of the Federal Republic of Germany. Here the Digital Landscape Model (ATKIS DLM 25/1) is used, which contains objects that correspond to the contents of the TK25.

In ATKIS the surface of the earth is divided into objects that are represented as points, lines and areas. The definition of the objects is described in a hierarchically structured feature catalogue. On top of the hierarchy the objects are divided into domains of object classes, e.g. hydrography, transportation, and vegetation. These domains are distinguished by groups, e.g. road traffic, and rail traffic, which are finally divided into object classes, e.g. road, and path. Each object is assigned to exactly one object class. A more detailed description of the objects is realized by attributes. For instance, an object of the object class road possesses the attribute motorway or federal road.

Depending on the topographic structure an object is divided into one or more parts. Attributes can be assigned to the whole object or to its parts. The geometric information (point, line or chain, area) is attached to the object parts. Topological relationships like the connection between nodes and edges of the road net are not stored explicitly, but the rules for building objects and object parts ensure that they can be obtained by identical nodes, e.g. crossroads, or lines, e.g. confluence of rivers which are stored as areas. Another spatial relationship the overpass/underpass reference is attached explicitly to an object part that is above or below another object part.

4.2 Integrating ATKIS Data

The ATKIS data is stored in the GIS Software Sicad/open. A simple query language has been developed to enable query operations on the ATKIS data. Possible queries are: selection of objects by their properties within a search area, description of attributes of a selected object, description of the geometry of a selected object. It is also possible to query spatial relationships of an object, e.g. topological connections.

The query language is implemented in Tcl/Tk. Hence it can easily be integrated into AIDA.

4.3 Representing ATKIS Data in a Semantic Net

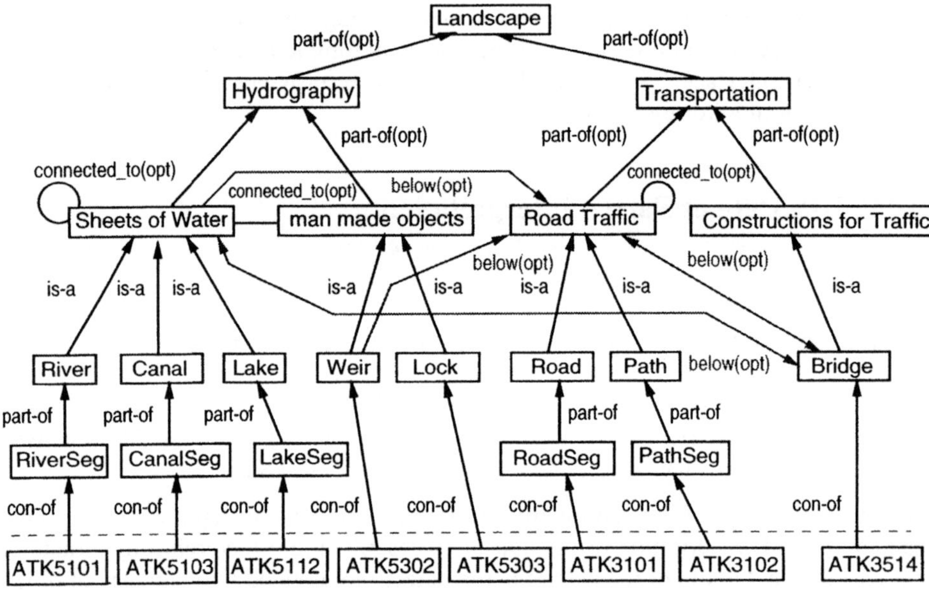

Fig. 3: Representation of ATKIS in a semantic net

Figure 3 shows a part of the semantic net representing ATKIS. In this part the *landscape* is divided into *hydrography* and *transportation*. Because both of them are not obligatory parts of the described scene, they are represented as optional parts of the landscape. *Hydrography* consist of the optional parts *Sheets of Water* and *man made objects. Sheets of Water* are specialized as *River*, *Canal* and *Lake*. This is represented by the *is-a* relationship. Each of this objects consists of at least one segment. These segments have a concrete realization as ATKIS objects. The transition from the abstract scene description to the concrete realization is represented by the *con-of* link. The number of the objects, e.g. ATK5101, correspond to the code of the object classes in ATKIS. The objects *Weir*, *Lock* and *Bridge* are not divided into segments because they have a direct representation in ATKIS.

In addition to the hierarchical relationships *attributed relations* are attached to the objects to model spatial relationships. For example a topological connection of objects is represented by the *connected-to* link. Objects that are on top or underneath another are connected by the *above* or *below* relationship.

5 Interpretation of Aerial Images

During the last years several methods for the analysis and interpretation of aerial images and other remote sensing data have been suggested.

A standard method is the analysis by classification (Mueller, 1992; Dennert Moeller, 1983). The limits of the classification method depend on the used data and are examined in (Michaelis, 1989). The use of different sensors by sensor fusion made a further improvement of the analysis. A good overview is given by Mueller (1989).

An interpretation of aerial images by knowledge-based object recognition was realized by Xu (1993) and led to a higher amount of recognized objects. All these methods contribute to the presented approach.

In the following the used method for the interpretation of aerial images is described which employs semantic nets and exploits an ATKIS database. As mentioned above the objects from the ATKIS DLM are used as a partial interpretation. Because the used DLM may be out-dated and has a limited accuracy, at first a verification of the ATKIS DLM with the used aerial images is necessary.

Hence the verification of the used ATKIS data is both the basis for the interpretation and the first application. The used domains of object classes are vegetation and hydrography. In the following this verification is described. The description is devided into two parts: The generation of an ATKIS scene description and its verification.

5.1 Generation of an ATKIS Based Scene Description

An ATKIS based scene description has to be generated for those areas of the aerial images which have to be examined. To do this the examined objects are taken from the ATKIS database and represented as instances in the semantic net, which is the basis of the following interpretation. One of the examined object classes is forest.

To generate a scene description a generic model, i.e. a concept net, is needed. The concept net consists of concept nodes and contains the considered objects in all needed levels of abstraction with all required relations between them. An instance net is built by instantiation of the concept net. This instance net is the realization of the concept net for the considered scene and contains the scene description, at first the ATKIS based scene description.

The used concept net is structurally a simplification of the net that describes the ATKIS database. Only the objects to be examined are included. The described spatial relations will be considered later and are not included yet. Figure 4 shows on the left side the concept net for the generation of an ATKIS based scene description for forest-objects. Two levels of abstraction are distinguished, the scene layer and the ATKIS layer.

Instantiation. To show how an instantiation process proceeds in the following the instantiation of the concept net in figure 4 is described. It starts with a seed node in the instance net. According to the strategy and its priority of rules the instantiation proceeds in a particular order along the relations postulated in the concept net, until no more rules can be applied and the instance net is complete.

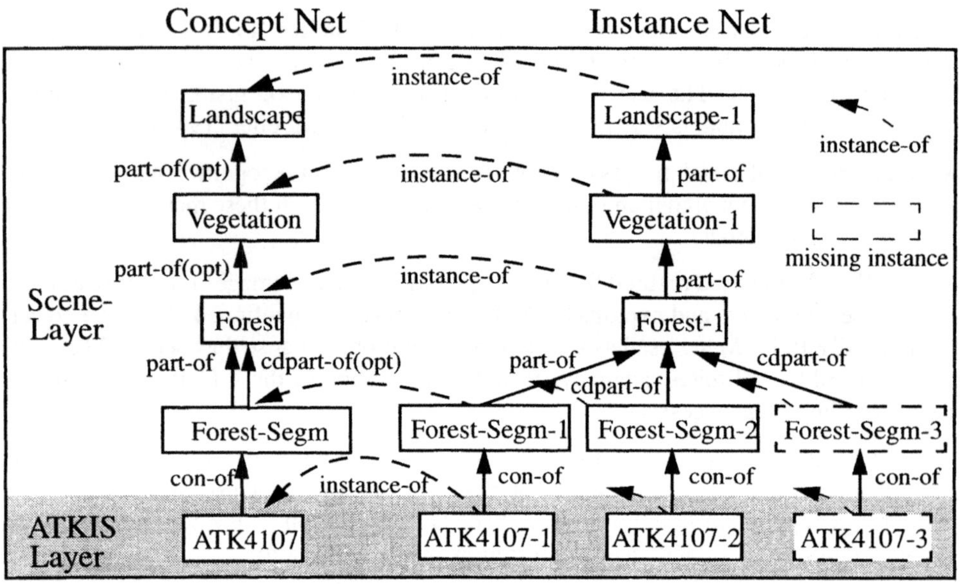

Fig. 4: Instantiation of a concept net

Here, to start the instantiation process a hypothesis I_H(*Forest*) is instantiated. That means a new node with the status hypothesis is created (*Forest-1* in fig. 4) and connected by an *instance-of* relation to the concept node C(*Forest*) in the concept net. The condition of the rule for complete instantiation (see 3.1) is not fulfilled because the obligatory parts are not present. Hence the rule for top-down propagation of a hypothesis is executed. This leads to an instantiation of the concept node C(*Forest-Segm*) along the *part-of* relation. The node *Forest-Segm-1* with the status hypothesis is created. In the same way *ATK4107* (4107 is the code of forest-objects in ATKIS) is instantiated as the hypothesis *ATK4107-1*. Now the bottom layer is reached. The concept node *ATK4107* at this layer, named ATKIS layer, contains a method that selects an ATKIS object 4107 through the ATKIS interface (see 4.2), which is inside the analysed area. If the method returns an ATKIS object the hypothesis becomes a complete instance. That means the instance node *ATK4107-1* changes its status from hypothesis to complete instance. Now all obligatory context independent parts of *Forest-Segm-1* are present. Hence this node becomes a partial instance.

The status partial instance blocks the bottom-up propagation till all obligatory and optional context dependent parts of *Forest-1* are detected. In this way all *Forest-Segm* of the *Forest*-1 are selected consecutively from the ATKIS database and are instantiated in the semantic net.

When all *Forest-Segm* of the *Forest* are extracted, the method returns no further object from the ATKIS database. Hence *ATK4107-3* becomes a missing instance instead of a complete instance. Consecutively *Forest-Segm-3* also becomes a missing instance. This is the condition to stop the creation of new *Forest-Segm* hypothesis via *cdpart-of(opt)*.

Finally the node *Forest-1* becomes a complete instance, because all obligatory parts are present.

Because no more top-down rules can be executed now the nodes *Vegetation* and *Landscape* are completely instantiated bottom-up. After this the net is completely instantiated.

In contrast to the obligatory parts the instantiation of optional parts can be performed as often as desired or not all. This has important consequences. The absence of a forest in the scene resulting in a missing instance of *Forest-1* does not lead to a missing instance of the node *Vegetation-1*.

ATKIS Attributes. In the instance net each forest possesses at least one *Forest-Segm* and one *ATK4107* node. The position and the form of the ATKIS objects are stored as attributes in data nodes. The data nodes are connected via *data-of* relations to the *ATK4107* nodes. This data nodes also contain other ATKIS attributes, which can be used in the later interpretation. For example, each forest object contains the ATKIS attribute vegetation feature. The vegetation feature contains the information whether a forest segment belongs to the group coniferous forest, deciduous forest or mixed forest. The stored attributes can be propagated from node to node.

5.2 Verification of the ATKIS based scene description

Now all needed ATKIS objects are instantiated in the semantic net. To continue with the verification in the aerial image, an aerial image layer has to be added to the concept net. As shown in figure 5 the aerial image layer is connected like the other sensor layers through a geometry layer to the scene layer. The addition of the new layers requires to reset the status of the nodes in the scene layer to hypothesis instance to avoid contradiction in the semantic net.

The verification in the aerial image uses the generated ATKIS based scene description as a hypothesis. Since the position and the form of each object is known from ATKIS, the

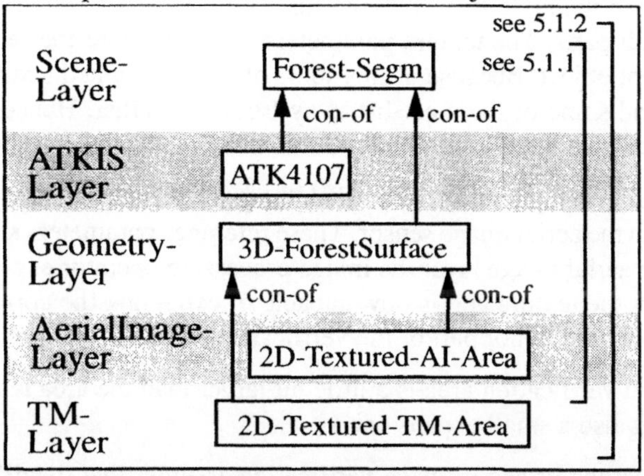

Fig. 5: Connection of sensor layers

hypothesis is propagated top-down to the aerial image where the corresponding object is expected in the same position with the same form as in ATKIS.

The verification of the semantic net in the aerial images leads to the instantiation of the hypothesis *3D-ForestSurface* and *2D-Textured-AI-Area*. This has to be verified in the aerial image. For this the node *2D-Textured-AI-Area* contains an image processing operator, which is executed after the instantiation of the hypothesis. In the following this operator is described

Verification in the Aerial Image. In the verification process the operator uses the prior knowledge of ATKIS, that there is an object of a particular type. The geographic coordinates of the objects are included in the instance nodes of the ATKIS layer as attributes. In the sensor layer the coordinates are converted into the aerial image coordinates. Exactly the area, which is obtained from ATKIS, is the area to be examined in the aerial image and is termed verification area in the following. The assumption of the verification in aerial images is, that the major part of the verification area is homogenous regarding particular texture parameters. Hence the texture parameter values for the searched objects show up as a maximum in the histogram of the texture parameters inside verification area. The searched objects can be extracted easily.

Problems occur if the assumption above is not fulfilled and the searched object does not fill the major part of the verification area. This happens if, for example, a forest area has been cleared and replaced by a housing estate or a field. In that case the texture parameter values of the searched object would result into a local maximum and the extraction of the object with the absolute maximum would lead to a falsification.

Hence at first a learning phase is executed. In this phase the most frequent texture parameters inside the verification area are determined, even if some of them belong to other object classes. The texture parameters, determined by the image processing operator, are returned to the corresponding nodes and saved as attributes.

These determined texture parameters are now propagated bottom-up from node to node until they merge in the node *Forest-1* (see fig. 4). In this node reference parameters are estimated from all parts. The texture parameters describe either the searched object or a complete different object. Because of this some of the determined texture parameters are close together and some of them are far away from each other. Hence the estimation of the reference parameters has to be robust (Press and Flanery, 1989).

The estimated reference parameters establish the prior knowledge how the scene objects are represented in the aerial image sensor. These reference parameters are now propagated top-down. In the aerial image layer the image processing operator exploits the additional knowledge about the texture and is now capable of extracting the correct object, even if the object fills only the minor part of the verification area.

The actually used verification area is a little bit larger than the area indicated by ATKIS in order to detect also a small expansion.

The described method of aerial image verification assumes that the differences between the examined objects in the aerial image data and the ATKIS data are not too large.

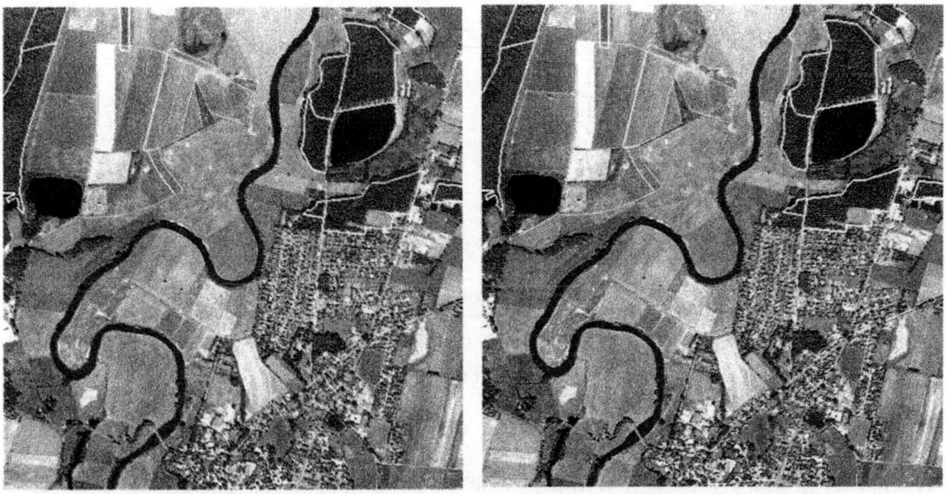

ATKIS Data Verified Data

Fig. 6: Result of verification

Otherwise the determined texture parameters would be to inaccurate causing false reference parameters.

On the other hand this kind of verification works without further prior knowledge about the object texture in the aerial images. This knowledge is acquired during the verification.

The advantage of this verification is the independence from contrast and brightness of the used aerial images. The aerial images could even be replaced by images of other sensors.

Verification Result. The result of the described verification of ATKIS data for the object class forest is shown in figure 6. The used texture parameters are the local variance and the local contrast in an area of $400m^2$. The left picture shows the objects taken from ATKIS before the verification. The right picture depicts the verified data. Inside the verification area all boundaries are adapted very good. One segment with texture features different from the other is not identified as forest.

5.3 Sensor Fusion

After the verification the actual position and form of the ATKIS objects are known. To improve the interpretation further and to make it more robust, more information can be taken from other sensor data, if available. The new sensors have to be included in the concept net, like the Landsat/TM-sensor in figure 5. The texture parameters for each verified forest segment are computed from the TM bands and stored as attributes.

Consecutively the interpretation uses this additional knowledge to improve the verification of objects.

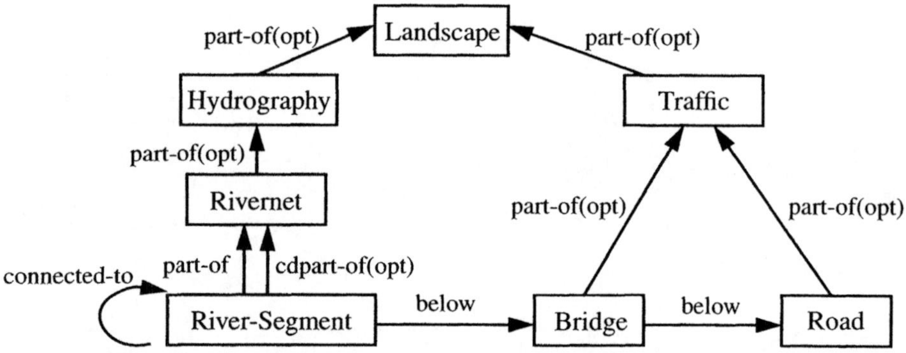

Fig. 7: Concept net with spatial relations

5.4 Spatial Reasoning

The use of spatial relations will improve the interpretation because additional knowledge is used. In the following an approach to find bridges above rivers is presented.

Figure 7 shows a concept net with spatial relations. The spatial relations are realized as *attributed relations*.

The prior knowledge that the river net and also the road net are contiguous is modelled by the *attributed relation connected-to*. That means each *River-Segment* is connected to another. Furthermore the prior knowledge, that a *River* runs only below a *Bridge* and that a *Bridge* only runs below a *Road*, is modelled by the *attributed relation below*.

During interpretation all *River-Segments* are checked for a break. In that case the hypothesis is instantiated that there is a *Bridge* above the *River*. After that the hypothesis for a *Road* above the *Bridge* is instantiated. To verify this hypothesis the aerial image is analysed in the interrupted area regarding texture and grey values. The result of this analysis leads to a verification or falsification of the *Road* hypothesis inside the break. After verification the *Bridge* becomes a complete instance.

To continue the interpretation from this point it is possible to find the contiguous road net by tracking the roads beginning with the found bridge.

6 Interpretation of Maps

The maps to be interpreted are taken from the German official map series 1:25000 (TK25). This map series was produced for the first time about one hundred years ago for the former German Reich. Therefore it is an important source for the detection of long term land use changes.

As data origin digital raster data of colour separated layers is used. The data is vectorized and attributes like line width and layer are attached. The vector data is then stored in a GIS. The interpretation process uses these pre-segmented primitives as basis data.

6.1 A Model of the Map Graphics

Figure 8 shows a part of the map graphic represented in a semantic net. The scene layer corresponds to that introduced in section 4.3. The *con-of* links of the *CanalSeg* represent the transition from the scene layer to the realization of a canal segment in ATKIS and in the map. The concept *CanalSegMap* represents the canal segment in the map layer. The depiction of a canal in the TK25 map depends on its width. For that reason a canal is specialized in a canal with a width less than three meters, which is depicted as a single blue line and a canal that is wider than 3 meters. This specialization is represented by the *is-a* relationship. A wide canal consists of two *banks* and the *canal bed.* In the map the bed is depicted as a blue area which is represented by a *con-of* link to the concept *blue area* in the graphics layer. The bank symbol can either be a *wall* or a normal *bank line* that is depicted as a *black line* or a *blue line.*

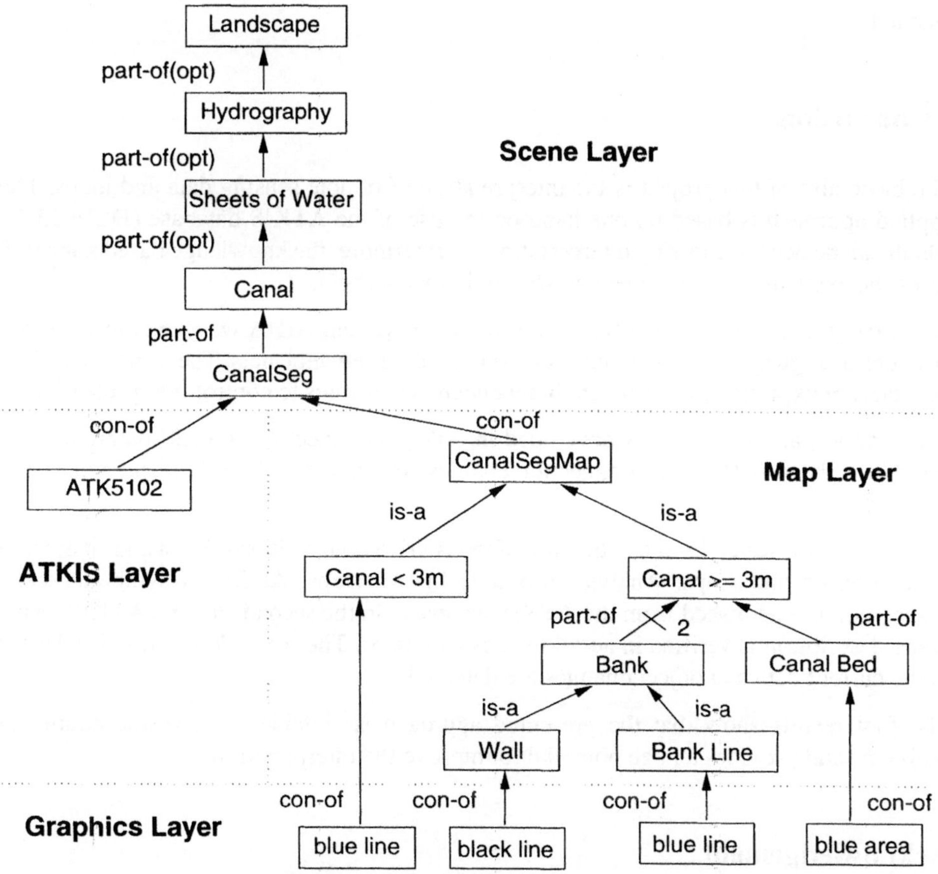

Fig. 8: Model of the map graphics

6.2 Instantiation

The essential process of interpreting the map corresponds to that described in section 5.1 and 5.2. At first the concept nodes of the scene layer are instantiated using the ATKIS data. After that they are verified in the map data using the instantiated scene description as a hypothesis. The hypothesis, e.g. of a canal segment, is propagated top-down. The relevant information obtained from ATKIS is also propagated top-down in the corresponding attributes of the nodes. Getting to the graphics layer the methods of the nodes *blue line*, *blue area* etc. are used to select corresponding map primitives. Here, the method is a GIS operation that creates a buffer around the reference geometry derived from ATKIS and searches within this buffer for primitives with the postulated attributes. The selected primitives are rated by a distance function. For each selected primitive an instance is created and the primitive is marked as used. According to the search strategy rules for creating partial and complete instances are then applied until the search goal is reached.

7 Conclusion

The basic aim of this project is the interpretation of remote sensing data and maps. The applied approach is based on one hand on the use of the ATKIS database (DLM 25/1), which can be accessed during interpretation. Furthermore, the knowledge based approach eases the extension to new object classes and other sensors.

To realize this the knowledge based interpretation system AIDA was developed, which provides a language for image interpretation based on semantic nets. The semantics of the network are exploited by a problem independent set of rules to control interpretation.

Furthermore, an interface was developed in order to connect the ATKIS database fully automatically to AIDA. This connection enables access to any ATKIS object and to any spatial relation.

The system was tested for the extraction of the ATKIS object classes forest and river from aerial images and maps. Analysis proceeds in two steps. At first a symbolic scene description is established from the ATKIS database. In the second step the ATKIS based scene description is verified in aerial images and maps. The scene description is adapted to the current data and object changes are detected.

The first results show that the presented approach is suitable to solve the examined problems and provides a huge potential to improve the interpretation.

Acknowledgement

This project is part of the research project "Semantic Modelling" funded by the DFG.

References

Braun, C., Kolbe. Th., Lang, F., Schickler, W., Steinhage, V. , Cremers, A.B., Förstner, W., Plumer, L. (1995), "Models for Photogrammetric Building Reconstruction", *Comput. & Graphics* , Vol. 19, No. 1.

Clement, V., Giraudon, G., Houzelle, S. , Sadakly, F. (1993), "Interpretation of Remotely Sensed Images in a Context of Multisensor Fusion Using a Multispecialist Architecture", *IEEE Trans. on Geoscience and remote Sensing*, Vol. 31, No 4, July 1993.

Dennert-Moeller, E. (1983), "Untersuchungen zur digitalen multispektralen Klassifizierung von Fernerkundungsaufnahmen mit Beispielen aus Wattgebieten der deutschen Nordseekueste", *PhD Thesis*, University of Hannover.

Foresti, G.L., Murino, V., Regazzoni, C.S., Vernazza, G. (1993), "Distributed spatial reasoning for multisensory image interpretation", *Signal Processing*, 32, pp. 217-255.

Grünreich, D. (1992), "ATKIS - A Topographic Information System as a Basis for a GIS and Digital Cartography in West Germany", *Geol. Jb.* Vol. A122, pp. 207-215, Hannover.

Haralick, R. M., Shapiro, G. S. (1993), "Computer and Robot Vision", *Vol 2, Addison-Wesley*, New York.

McKeown, et al. (1985), "Rule-Based Interpretation of Aerial Imagery", *IEEE Trans. on Pattern Analysis and Machine Intelligence*, Vol. PAMI-7, No. 5, pp. 570-585, Sept. 1985.

Matsuyama, T., Hwang, V.S.-S. (1990), "SIGMA : A Knowledge-Based Aerial Image Understanding System", *Plenum Press*, New York.

Mayer, H., Steger, C. (1996), "A New Approach for Line Extraction and its Integration in a Multi-Scale, Multi-Abstraction-Level Road Extraction System", *IAPR TC 7 Workshop: Methods for Extraction and Mapping Buildings, Roads and other Man-Made Structures from Images.*

Michaelis, M. (1989), "Untersuchungen zum topographischen Potential von Daten des Erderkundungssatelliten SPOT mit Hilfe digitaler Bildverarbeitungsmethoden", *PhD Thesis*, University of Hannover.

Mueller, W. (1989), "Verknuepfung von Radarbildern mit Bildern optischer Sensoren in einer geokodierten Datenbank", *final report of the research project*, University of Hannover.

Mueller, W. (1992), "Auswerteverfahren und exemplarische Auswertung photographischer Aufnahmen aus dem Weltraum", *PhD Thesis*, University of Hannover.

Niemann, H., Sagerer, G., Schröder, S., Kummert, F. (1990), "ERNEST: A Semantic Network System for Pattern Understanding", *IEEE Trans. on Pattern Analysis and Machine Intelligence*, Vol. 12, No. 9, pp. 883-905, Sept. 1990.

Press, W., Flannery, B. (1989), "Numerical Recipes, The Art of Scientific Computing", *Cambridge Univ. Press.*

Shapiro, S. C. (Ed.) (1992), "Encyclopedia of Artificial Intelligence", *John Wiley & Sons*, New York.

Xu, Y. (1993), "Contextimage, ein objektorientiertes System zur wissensbasierten Bildanalyse und Objekterkennung mit Anwendung in der Photogrammetrie", *PhD Thesis*, University of Hannover.

Semanitic Modeling for the Acquisition of Topografic
Information from Images and Maps
ed. by W. Förstner and L. Plümer

A new approach for satellite image analysis by means of a semantic network

Dietmar Kunz, Klaus-Jürgen Schilling and Thomas Vögtle

Institut für Photogrammetrie und Fernerkundung, Universität Karlsruhe
e-mail: kunz@ipf.bau-verm.uni-karlsruhe.de, schi@ipf.bau-verm.uni-karlsruhe.de,
voegtle@ipf.bau-verm.uni-karlsruhe.de
http://www-ipf.bau-verm.uni-karlsruhe.de/

Abstract

This paper investigates synergy effects between satellite images and a digital topographic database. The use of such topographic databases, which are actually built up in many countries (e.g. ATKIS-DLM200 data in Germany), can support the satellite image analysis. This digital database offers a geometric as well as a semantic prediction for every object in the satellite image. Therefore, an automated, knowledge based feature extraction can be performed. Because of the relatively poor geometric resolution of common satellite sensors the project will concentrate on main area-based landuse classes like 'settlement', 'forest', 'water' or 'agriculture'. The result of this feature extraction is a symbolic description of the image content.

Both, the symbolic description and the digital database, are transfered in a semantic network, a compact formalism for structuring the entire knowledge. Beside this, the semantic network is used for verification and classification of the above mentioned objects. Therefore, suitable valuation and analysis functions have to be created. In the last step, a change detection between satellite image and database information will lead to an update process of the topographic database.

1 Conception

This paper is based on a project to investigate synergy effects between satellite image analysis and digital database processing. The aim of this approach is an improved and automated image analysis concerning the determination of land use classes by means of specific knowledge, which is represented in digital topographic databases. On the other hand, these more actual results of the image analysis process in comparison with a digital database can also be used for updating this digital database.

Since a couple of years a lot of countries all over the world have been building up digital topographic databases for planning, modelling and visualization purposes. In Germany, the surveying administration has designed and created ATKIS (Amtliches Topographisch-Kartographisches Informationssystem) as such a topographic database. ATKIS is divided in three different landscape models DLM25, DLM200 and DLM1000, which contain mainly the objects of the official topographical maps in scales 1 : 25 000, 1 : 200 000 and 1 : 1 000 000 respectively. Because of the poor resolution of common satellite sensors (approx. 10m to 30m) this project will concentrate on DLM200 information.

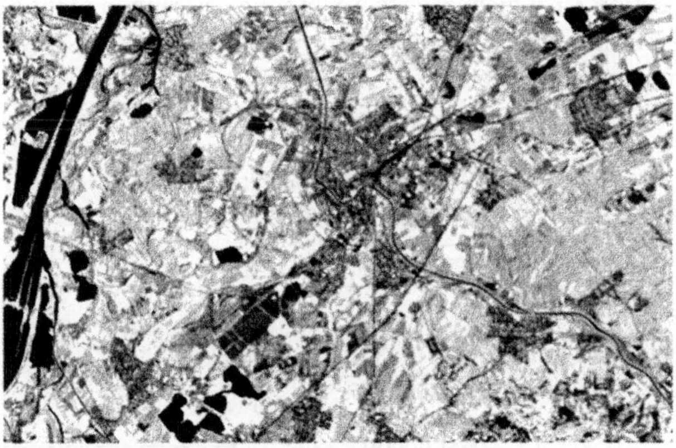

(a) Satellite image information (Landsat TM, channel 5/4/2, 27.04.1993, approx. 15km x 12km)

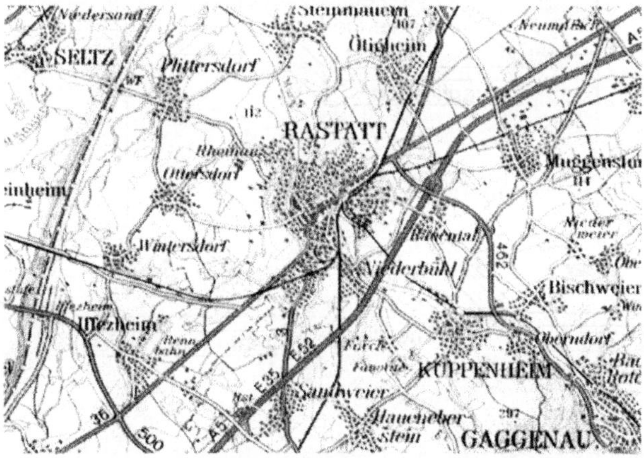

(b) Topographic database 'ATKIS-DLM200' (same area as (a))

Fig. 1: Input data

This digital ATKIS-DLM information offers a-priori a complete and area-wide semantic description of the content of the satellite images (see Fig. 1(a) and 1(b)). Therefore, a model-driven *top-down* approach can be integrated in the commonly data-driven *bottom-up* process of satellite image analysis.

ATKIS-DLM200 allows a prediction of the semantic of all corresponding image segments, therefore, the actual manifestation of the object features (e.g. spectral signature, texture, shape, size etc.) can be 'learned' by robust estimation. This specific knowledge added by general knowledge about the scenario (e.g. sensor characteristics, reflectance behaviour etc.) is used to realize an improved, knowledge-based segmentation and feature extraction (see section 3), which leads to a symbolic description of the satellite image (see Fig. 2). This step from pixel level to symbolic description is necessary for comparison of the two different information sources.

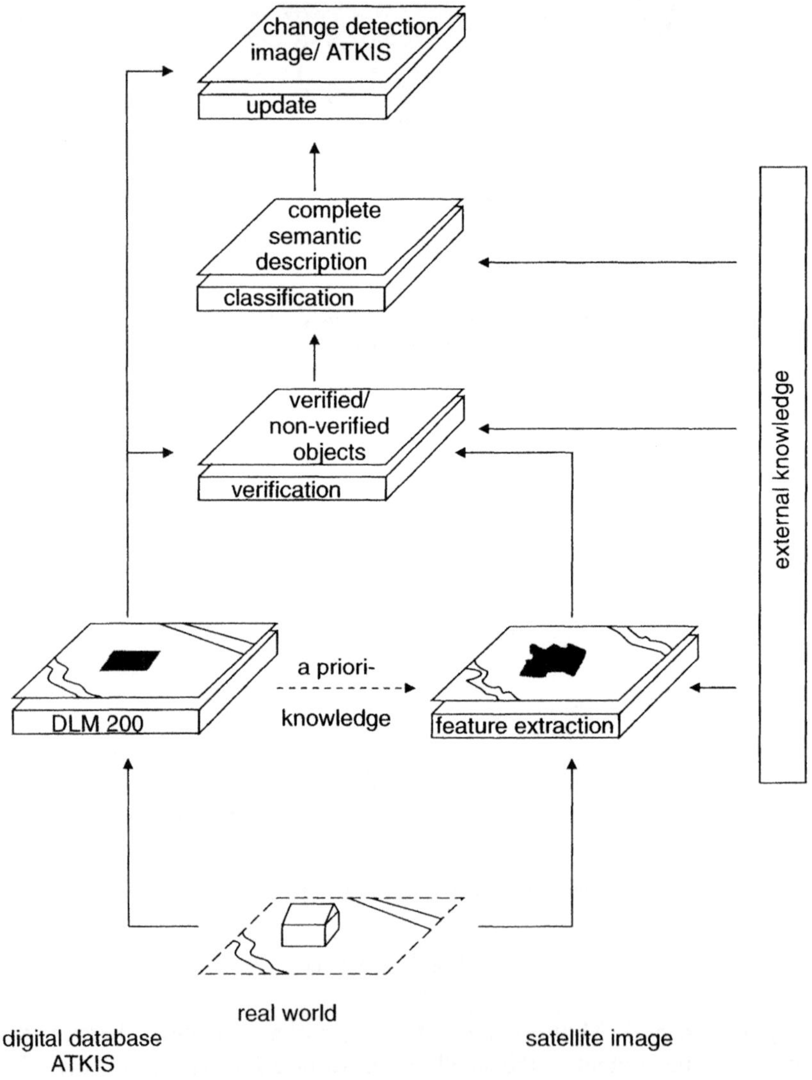

Fig. 2: Flowchart of the analysis procedure

The knowledge representation of these two information layers is based on *semantic networks*, which offer a compact formalism for structuring the entire knowledge of ATKIS-DLM information as well as the symbolic image description. Additionally semantic nctworks contain tools for valuation of features within the decision process. Thus semantic networks are used for knowledge representation <u>and</u> knowledge processing.

In this project, a semantic network is built up with the ATKIS-DLM information, which leads to a specialized scene description. With this automatically created semantic network the verification and classification process is carried out for the extracted image segments. Besides of knowledge representation, the valuation processes as well will be modelled in ERNEST.

The result of the classification process is a complete semantic description of the satellite image. In a future step a change detection for the update process will be performed between the semantic description of satellite image and ATKIS. This topic will not be discussed in this paper.

2 ATKIS

The digital database ATKIS contains objects, which are defined in a catalogue by the surveying administration (object type catalogue). This catalogue is structured hierachically in three levels (AdV 1989); (Harbeck 1994). At the highest hierachical level there are seven general classes ('fixed point', 'settlement', 'traffic', 'vegetation', 'water', 'relief' and 'area'). This classes are subdivided in specific subclasses and the subclasses are further divided in concrete ATKIS objects. For differentiation all objects and classes are coded by a unique number (4 digits). The first digit identifies the general class, the second delivers the subclass and the last two digits realize the concrete object. As an example the

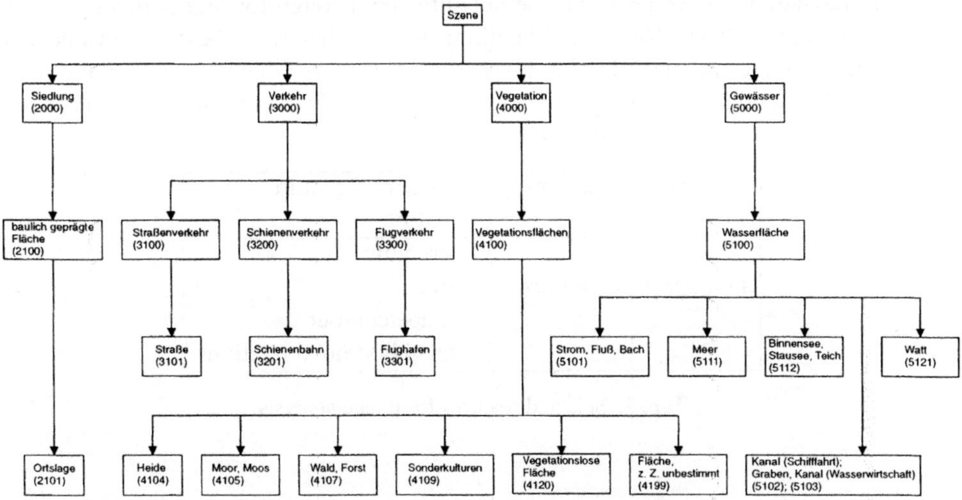

Fig. 3: ATKIS-Level of the semantic network

object number 4107 describes 'forest', which belongs to the subclass 'vegetation area' (number 4100) and the general class 'vegetation' (number 4000). ATKIS is divided in three landscape models (DLM25, DLM200 and DLM1000) related to the scales 1:25000, 1:200 000 and 1:1 000 000 respective. Each of these models contains geometry and semantic meaning of the objects of the official topographic maps in digital vector-oriented representation, created by digitizing these maps. Each layer represents one object class like 'streets', 'settlement', 'forest' or 'water'. Because the digital database ATKIS was a definition of the german administration for a topographic information system, only german terms are used for the description of the ATKIS objects and classes in this paper (see Fig. 3). In this project only objects can be used, which are represented in the image information (e.g. no line-oriented classes like streets, railways, small-rivers or objects like frontiers or sights) Therefore, a subset of the original object type catalogue has to be built. The selected objects from the catalogue are the basis for the realization of the semantic network in section 4.3.

3 Feature Extraction

3.1 Selection of relevant features

Before selecting appropriate object features, we have to define the objects which should be extracted by image analysis and updated in the digital database. With respect to the poor geometrical resolution of common satellite sensor systems (pixel size on ground about 10 m to 30 m) we concentrate on main area-based landuse classes like 'settlement', 'water', 'forest' or 'agricultural areas'. As topographic database we use ATKIS-DLM200 (see section 2). In the future, if sensors with higher resolution will be available, the DLM25 (1 : 25 000) should as well be introduced in the analysis process.

The common features in satellite image analysis, i.e. the spectral signatures (mean values, standard deviations), have proved to be not sufficient enough for higher quality results (Bähr and Vögtle 1991), (Vögtle and Schilling 1995). Therefore, these features have to be extended to spectral as well as non-spectral parameters which can contribute to an improved distinction between the defined object classes. Thus we also take into account geometrical and structural features (Tab. 1):

spectral features	spectral signature texture
non-spectral features	structure size shape/contour neighbourhood relations

Tab. 1: Selected features for image analysis

3.2 Automation in feature extraction

The automated extraction of the above defined features is based on the a-priori knowledge represented in the topographic database ATKIS-DLM200, which offers both a (possibly not actual) geometric and a semantic description of those land-use objects to be extracted from satellite images. In opposite to the commonly used method where a human operator interactively has to define some representative training areas based on his experience and intuition, now *all* (DLM-)objects of the same class within the geocoded satellite image can be used as training areas without human interaction. Therefore, we normally get a very large sample and we have to perform a robust estimation of the defined features to exclude disturbances caused by errors in the topographic database or in the image information, e.g. out-of-date status of some polygons (contour lines), digitizing errors or errors in the geometrical correction of the satellite image. For a robust estimation we have to assume that most of the objects we are interested in (at least more than 50%) have not changed, a pre-condition which is fulfilled in most cases.

The feature extraction process in this approach contains a hierarchical concept. The *spectral characteristic* of the objects is still one of the most important features in satellite image information. To get a robust estimation of the spectral signatures of each object class we have to extract only the *representative* reflectance values for this class. For relatively homogeneous objects like 'water' or 'forest' statistical methods have been proved to be sufficient, e.g. histogram analysis or a median estimation. Inhomogeneous objects like 'settlement areas' can not be treated in this way. Typically these areas contain a strong mixture of different (sub-)objects (man-made objects, meadows, gardens, trees, water areas etc.), and therefore, a wide range of reflectance values. Nevertheless, the accumulation of vegetation-free pixels caused by man-made objects (e.g. buildungs and traffic areas) can be seen as representative for 'settlement'. With respect to this knowledge we can extract the vegetation-free pixels by means of the NDVI (*Normalized Difference Vegetation Index*).

$$NDVI = \frac{(IR - R)}{(IR + R)}$$

IR – reflectance values in near infrared
R – reflectance values in visible red

Then the determination of the representative signature for 'settlement' can be performed by a statistical approach only based on these pixels. The next step in this hierarchical extraction process is the determination of the geometrical features 'size' and 'shape', which leads to the problem of finding object contours. Supposing that the appearance of the land-use objects in satellite images can by characterized by a significant accumulation of pixels with typical reflectance values compared to the surrounding areas, the spatial distance of valid pixels which belong to the same class, can be a criterion to find the object boundaries. We have investigated two different methods, *mathematical morphology* and *triangulation networks*. As an example Fig. 4(a) shows an original detail of a satellite image and Fig. 4(b) the significant vegetation-free pixels (in black) as candidates for the object class 'settlement'. It is obvious, that edge detection algorithms can not be successful in this case. Therefore, we tried to perform algorithms of mathematical morphology: In a first step iterative *dilation* close the gaps inside the objects, afterwards the same number of iterations of *erosion* restore the original object size. Caused by the dimension of such gaps it is difficult to predict the number of required iterations.

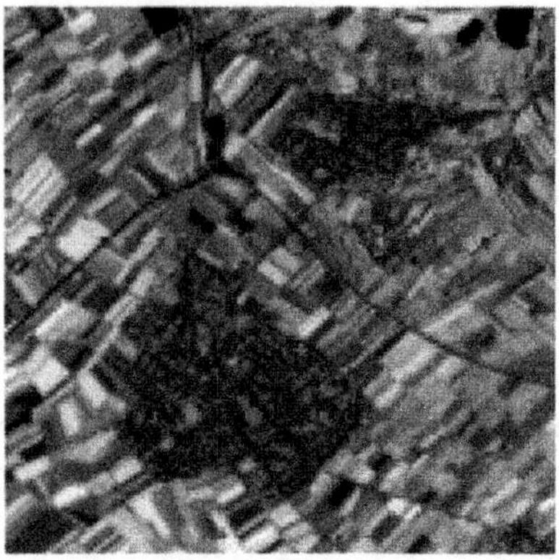

(a) Detail from test area 'Speyer' (approx. 3km x 3km)

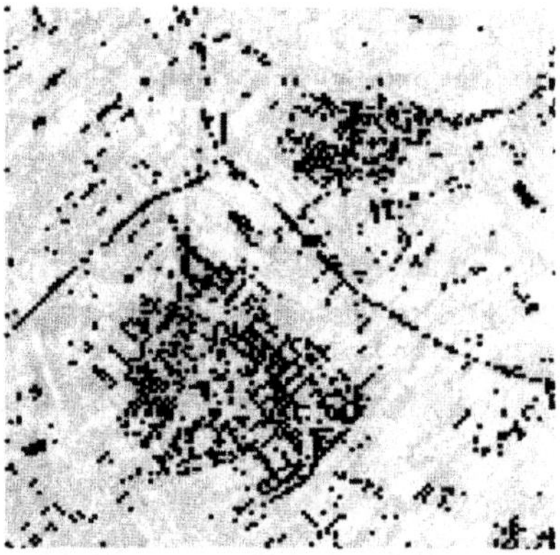

(b) Marked vegetationfree pixels

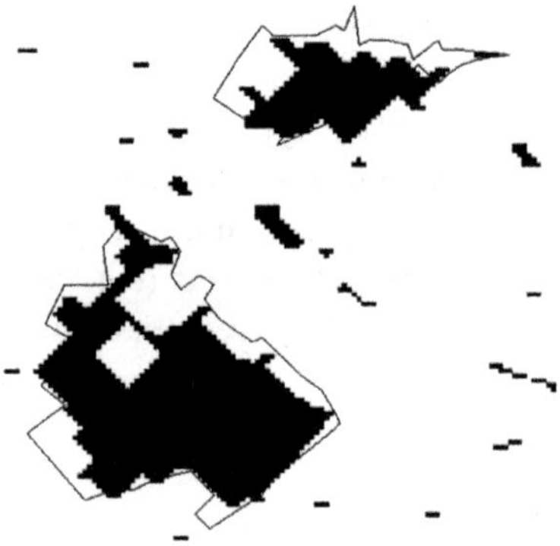

(c) Segmentation of image objects by morphology operators (black lines: orig. DLM200 object contours)

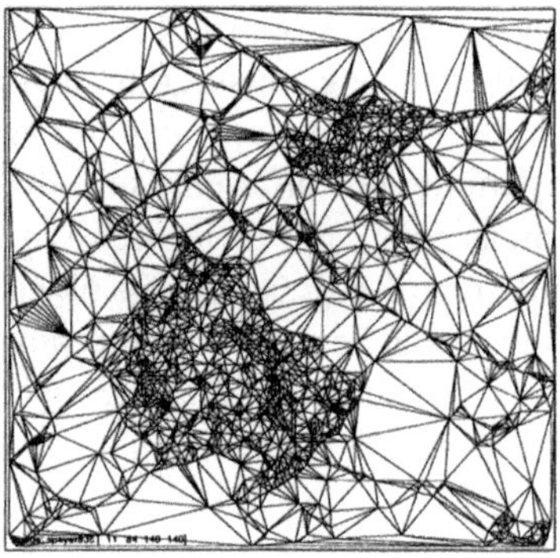

(d) Triangulation network using vegetationfree pixels (see Fig. 4(b))

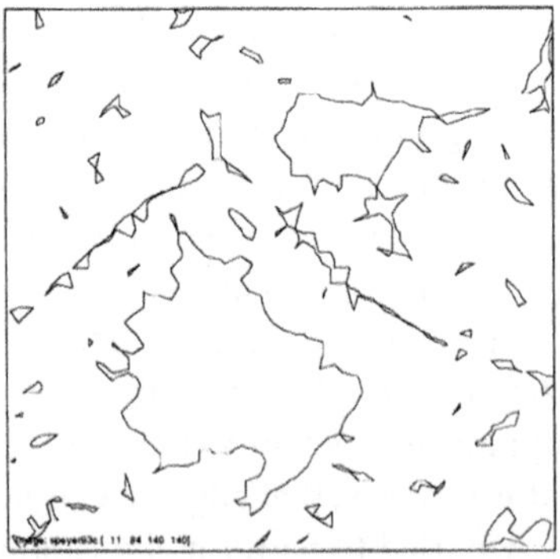

(e) Selection and fusion of valid triangles to segments

Fig. 4: Extraction of geometrical features 'size' and 'shape'

On the other hand a higher number of iterations results in a significant distortion of the object contour (Fig. 4(c)). A more successful method may be the use of triangulation networks, e.g. Delaunay triangulation (Fig. 4(d)) of the marked pixels of Fig. 4(b). Based on all DLM200 settlement objects in the satellite image, statistics of the triangle perimeters (mean value, standard deviations) can be determined. This is used as *structural parameter* related to the 'compactness' of the object.

On the other hand *valid* triangles can be extracted by means of this perimeter statistic. After a fusion of all adjacent valid triangles the object contours can be obtained (Fig. 4(e)). The same procedure can be applied to all other – also homogeneous – object classes.

The *size* of the objects is derived from the coordinates of the contour lines. The selection of suitable *shape* parameters has proved to be more difficult. Among a lot of parameters which have been investigated, 'compactness' and 'mean curvature' seem to be most suitable:

$$compactness = \frac{P^2}{4\pi A}$$

where P – perimeter of object contour
A – area of object

$$mean\ curvature = \frac{\sum_{i=1}^{n} |\Delta\alpha_i|}{n}$$

where $\Delta\alpha_i$ – change in direction at contour-point i

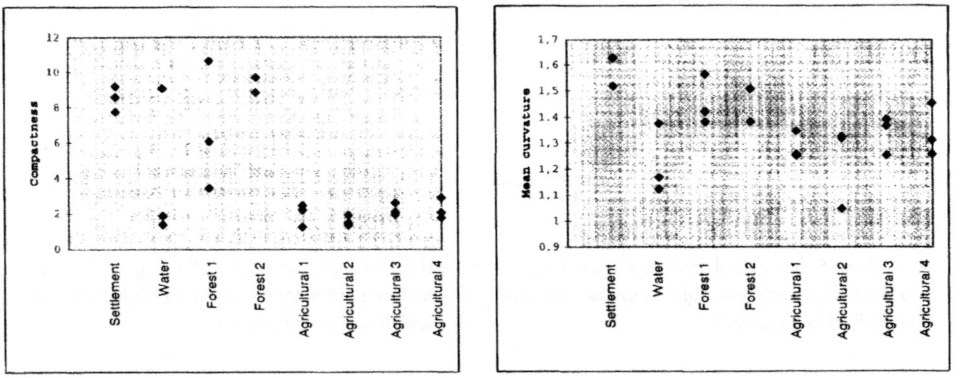

(a) (b)

Fig. 5: Shape parameters 'compactness' and 'mean curvature' for different training areas of different object classes

Fig. 5 shows these parameters for different training areas and different object classes.

In the last step of feature extraction we can calculate *texture* parameters inside the object contours. A very simple but efficient parameter is the standard deviation σ_g of the greyvalues g_i inside the objects. Because of some disturbances, for instance caused by errors in the topographic database or errors in the geometrical correction of the satellite image, only the main greyvalue distribution has to be extracted by a histogram analysis. The result is a *modified standard deviation* σ_g^*, which can be used as robust texture parameter (Fig. 6(a)).

Also test series with the well-known *Haralick parameters* (Haralick and Shapiro 1992a); (Haralick and Shapiro 1992b) have been carried out. They have shown that *homogeneity* is one of the best parameters to destinguish between the above defined object classes (Fig. 6(b)):

$$homogeneity = \sum_{j=1}^{n} \sum_{i=1}^{m} \frac{w(i,j)}{1 + (g(i) - g(j))^2}$$

where $w(i,j)$ – element of the well-known co-occurrence matrix
$g(i)$, $g(j)$ – greyvalue in a row / column

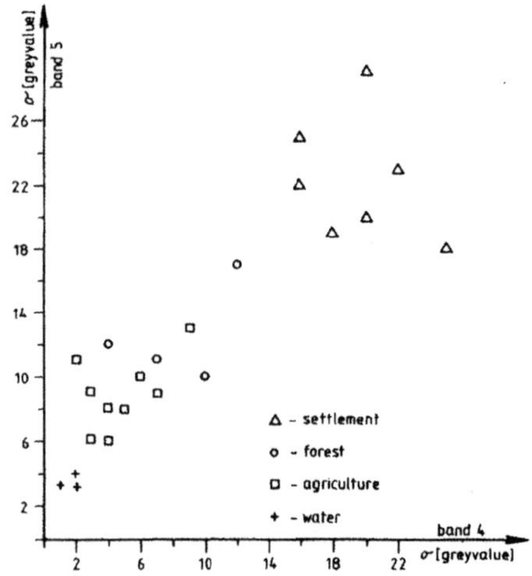

(a) Modified standard deviation as texture parameter for different object classes and different training areas

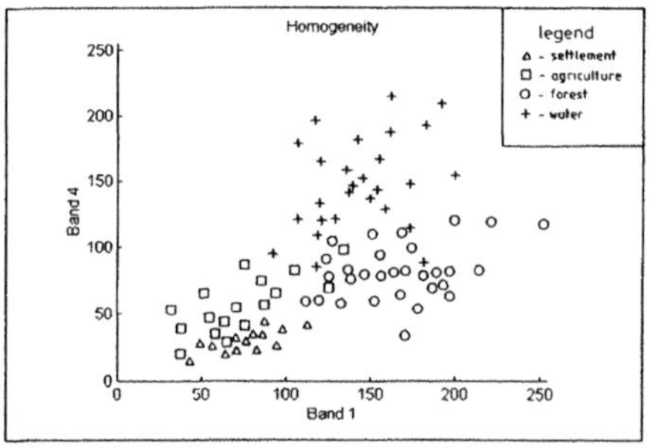

(b) Haralick parameter "homogeneity" as texture parameter for different object classes and different training areas

Fig. 6: Texture parameters 'homogeneity' and 'modified standard deviation'

Neigbourhood relations are determined by storing the identification code numbers of the adjacent segments to each object.

The result of the whole feature extraction process is a symbolic description of the image content by segments (contour lines) with a number of attributes (signature, structure, size, shape) and topological information (neighbourhood relations).

4 Semantic modelling

The first step in semantic modelling of the image content will be the *verification* of the ATKIS-DLM200 objects in the satellite image by means of the extracted candidates (segments). If objects have not changed and therefore have (with a certain probability) no significant difference, they can be verified. If not, a general classification procedure is applied to these non-verified objects (see Fig. 2). For the comparison of the two different information sources, it's necessary to introduce this methods on higher, symbolic level. For this special type of data processing general and specific knowledge about the topographic objects is used. Therefore, the analysis system must be able to represent and process this knowledge. Different formalisms for knowledge representation are known like *predicate logic, rule-based systems, formal grammatics* or *semantic networks* (Barr and Feigenbaum 1982); (Schefe 1986); (Winston 1987); (Niemann and Bunke 1987); (Lusti 1990).

4.1 Semantic network

Semantic networks belong to the most usable schemes concerning knowledge representation (Paulus 1992). The semantic networks so far have been used for speech recognition (Kummert 1992), industrial (Niemann et al. 1990b) and medical (Bunke 1985) applications and aerial image analysis (Quint and Sties 1995). The use of a semantic network for satellite image analysis is new.

A semantic network contains two different types of knowledge: *declarative* and *procedural knowledge*. Declarative knowledge consists of concepts and links, while procedural knowledge contains methods for determination of attributes of concepts as well as for valuation of concepts and relations.

The declarative knowledge can be described by a marked and directed graph, where the nodes are attributed. Therefore, graph-theory can be applied for the control and analysis process within a semantic network.

For using and modelling the knowledge of topographic database (ATKIS-DLM200) and actual image – as well as central control unit – ERNEST (Erlanger Semantisches Netzwerksystem) is applied in this project (Niemann et al. 1990a); (Kummert et al. 1993). In the semantic network system ERNEST there are three different types of links between two nodes: "part-of" (bst), "specialization-of" (spez) and "concrete-of" (kon). The links describe the relation between two nodes. The nodes represent various objects, events, ideas, or abstract concepts. For a general discussion of semantic networks, see e.g. (Quillian 1968); (Brachman 1977); (Woods 1975); (Findler 1979).

4.2 Model of the analysis process

This project is based on three models (see Fig. 7). Each model resides in his own domain. We distinguish between three domains: the scene domain, the image domain and the ATKIS domain. The scene domain is our environment, the real world. The ATKIS domain and the image domain comprise representations of the real world in ATKIS and image. The generative model contains domain independent, common sense knowledge about the real world. The generic models in ATKIS and image domain are specializations of the generative model and they reflect the particularities of the representations of the real world in digital database ATKIS and satellite image respectively. The analysis is based on a comparison of the two models in image and ATKIS domain. One of the main characteristics of the system is that ATKIS information is used to refine automatically the model used for image analysis.

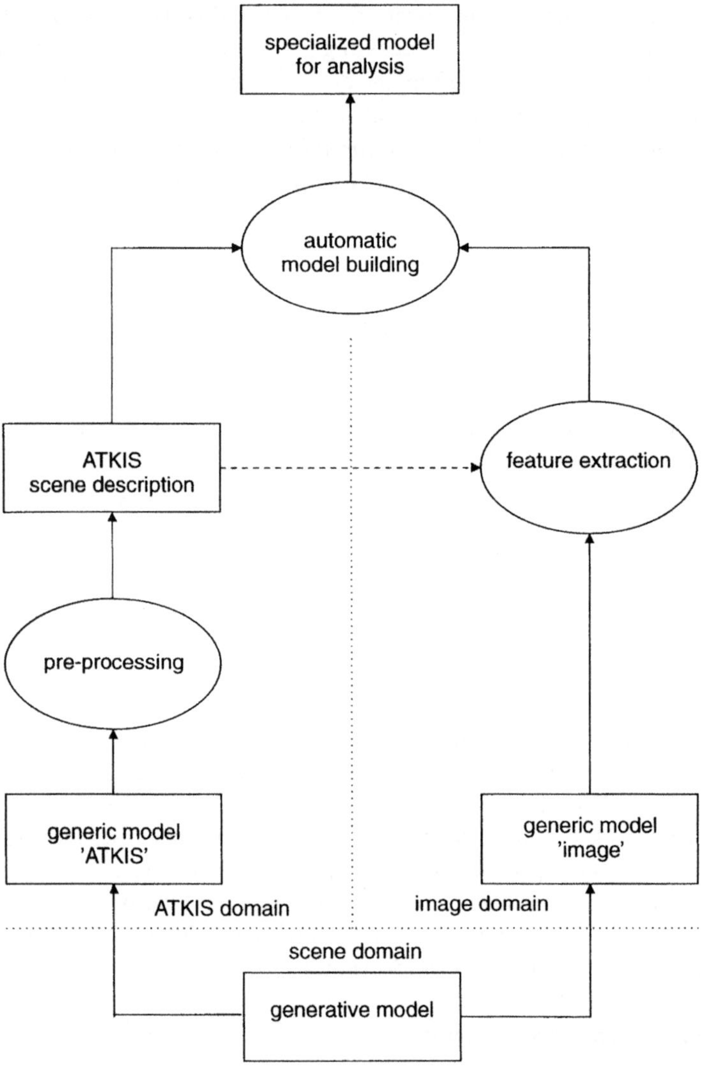

Fig. 7: Structure of the used models

The pre-processing in ATKIS domain gives us a complete description of the scene in the limits of the content of the digital database ATKIS. Using this description plus selected attributes from the feature extracted generic model in image domain, we obtain a specialized model of the concrete scene to be analyzed. The generic model in the image domain delivers the mean values of the parameters for an object class and the ATKIS information the geometry and the semantic meaning of a polygon. Thus, knowledge from the ATKIS domain is integrated in image analysis.

4.3 Realization of a semantic network

The specialized model for image analysis is transformed in a semantic network specialized in analysis of the underlying scene. This generated network is used for image analysis.

The semantic network, which is used in this project, is structured in four levels. As an example, Fig. 8 shows all four levels for the representation and analysis of settlement.

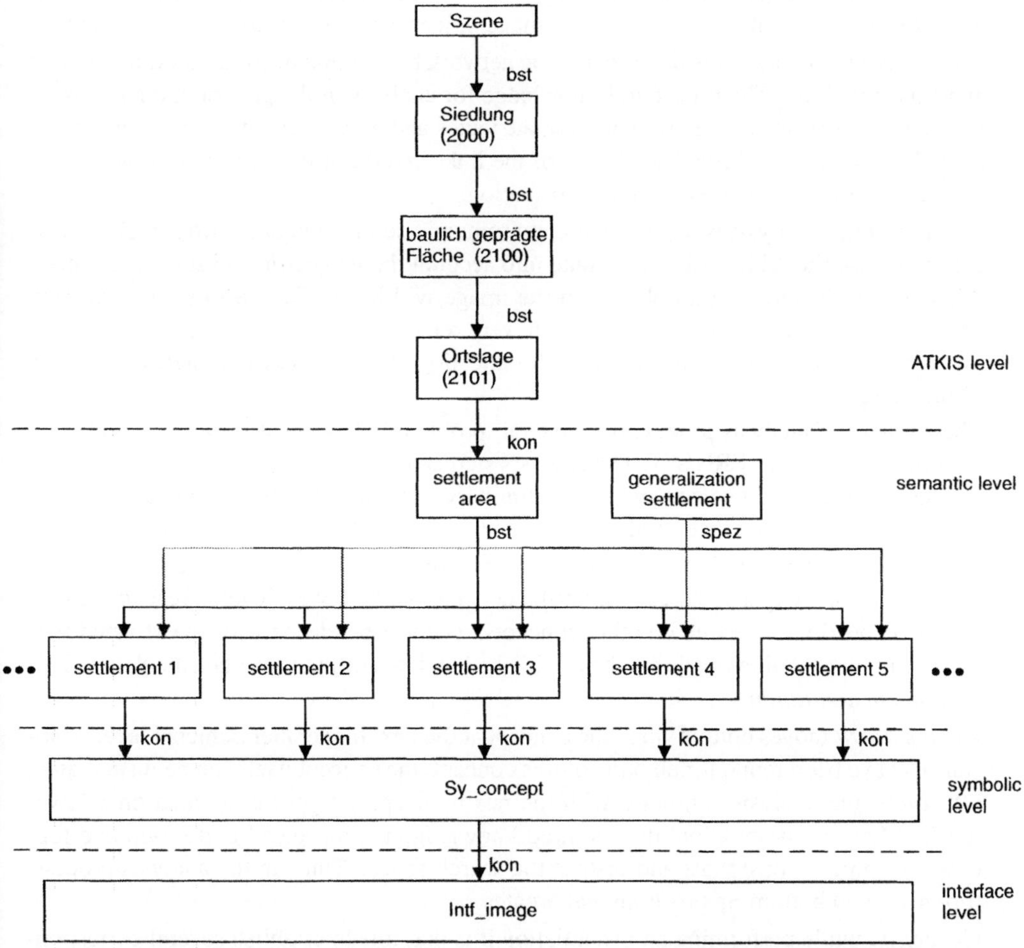

Fig. 8: Structure of the semantic network (e.g. for DLM class 'settlement')

On top there is the ATKIS level, which was described in section 2. The aim of our analysis is to divide the satellite image in the object types of ATKIS level.

Under the ATKIS level, the semantic-level is located. In this level the symbolic objects of the symbolic level will be assigned to a semantic meaning. Most of the analysis process is performed in the symbolic level and is explained in detail in section 4.4. The interface level introduces the extracted image segments to the analysis process within the semantic network.

4.4 Data Analysis

The main topic of analysis process will be the verification and classification of image segments (procedural knowledge processing)(Fig. 2). Beside of this image information – declarative knowledge realized by concepts and links – also the results of feature extraction (additional procedural knowledge) have to be introduced in the semantic network to enable a successful verification of all image objects based on 'learned' information.

This analysis is realized inside the semantic network by creation of suitable valuation- and analysis-functions. The procedural knowledge for analysis of the generic model 'image' consists of functions for determination of attributes and parameters of concepts as well as for valuation of the achieved attributes, of the links and the structure relations, where the database (DLM200) acts as a-priori information.

The aim of this analysis is to get a complete semantic description of the image. For verification we use the ATKIS objects to take into account the geometric and semantic aspect. After the verification not all objects in the image will be verified. So the classification process has to deal with the objects which were rejected during verification. As result of this combination of the verification and classification we will have a complete description of the image.

The strategy of analysis process is a general, problem independent strategy provided by the semantic network ERNEST. The analysis starts by creating a modified concept for the goal concept. A modified concept is a preliminary result and it reflects constraints for the concept that have been determinated out of the context of the current analysis state.

Following the hierarchy in the semantic network top-down, the concepts on lower hierarchical levels are expanded stepwise until a concept on the lowest level is reached. Since this concepts do not depend on other concepts, its correspondence with an extracted segment from the satellite image can be established and its attributes can be calculated. This is called *instantiation*.

Analysis now moves bottom-up to the concept at the next higher hierarchichal level. If instances have been found for all parts of this concept, the concept itself can be instantiated. Otherwise the analysis continues with the next concept not yet instantiated on a lower level. After an instantiation, the acquired knowledge is propagated bottom-up and top-down to impose constraints and restrict the search space. Thus, in the analysis process, top-down and bottom-up processing alternates.

Generally, while performing an instantiation it is possible to establish several correspondences between a concept and primitives in the ATKIS database. However, only one of these correspondences leads to the correct interpretation. Usually it is not possible to decide ultimatley at the lower levels which correspondence is correct, so all possible correspondences have to be taken into account.

Thus, the image analysis is a search process, which can be represented graphically by a tree structure. Each node of the tree represents a state of the analysis process. If several correspondences are possible in a given state, the search tree is splitted.

The analysis process continues with that node of the search tree which is considered to be the best according to a problem dependent evaluation. As it's known, the problem of finding an optimal path in a search tree can be solved by the A^*-algorithm.

Its application is possible if the path from the root node to the current node can be evaluated and if an estimation can be given for the valuation of the path from the current node to the terminal node not yet known containing the solution.

The functions which evaluate the states of the analysis are very important since they are not responsible for the efficiency of the search, but they are also decisive for the success or failure of the analysis. The valuation of the search path will be related to the valuation of the analysis goal. The valuation of the goal is calculated considering the valuations of the instances and modified concepts which will be created in the path from the current node to the solution node. When an instantiation is performed, implicitly a hypothesis of match is established between the concept under instantiation and the chosen primitives from the ATKIS database. Since we can not decide ultimately at the moment the instantiation is performed, if it is the correct one, a so called error-band is applied. At the level of each concept in the semantic network we have a dichotomous frame of discernment with the events: the chosen primitives *match* or *do not match* to the concept.

The valuations computed for the instances and modified concepts in each state of the analysis are measures of our subjective belief in these hypotheses. They take values between 0 and 1 and we interpret them as basic belief masses in the framework of the Dempster-Shafer-Theory of evidence (Shafer 1976); (Dempster 1967); (Dempster 1968); (Smets 1991). The higher a valuation, the stronger our subjective belief in the corresponding hypothesis. The different valuations are combined and propagated in the hierarchy of the semantic network to result in the valuation of the analysis goal.

We evaluate two aspects for our hypotheses of a match: the compatibility and the fidelity of the model. The compatibility evaluates an analysis state considering the principles of perceptual grouping. It is calculated based on geometrical, topological and radiometrical properties of the image primitives only.

The fidelity of the model determines the quality of fit between image primitives and specific model gained through the analysis of the DLM-information. The compatibility can be seen as a measure of the ability to form an object of the generic model with the chosen image primitives. Model fidelity is a measure of the ability to form exactly that object, which is predicated by ATKIS.

At the moment Dempster-Shafer-Theory is used for automated image analysis. In the following update process Fuzzy-Theories and classical statistical methods based on Bayes-decision-rules will be investigated.

5 Conclusion

Experiences with an extented feature base and a special segmentation confirm the efficiency of this concept by achieving a better separability of object classes. The systematic structuring of knowledge in semantic networks is a basic precondition for this integrated knowledge processing. But the creation of concrete semantic network has proved to be a very complex and time consuming task. Therefore, specific software has to be developed for automation of this procedure. The determination of suitable valuation functions in the decision process and the treatment of uncertainty is the most important aspect.

References

AdV (1989) *Amtliches Topographisch-Kartographisches Informationssystem (ATKIS)*, Arbeitsgemeinschaft der Vermessungsverwaltungen der Länder der Bundesrepublik Deutschland (AdV), Hannover.

Bähr H.-P., F. Quint, U. Stilla (1995) *Modellbasierte Verfahren der Luftbildanalyse zur Kartenfortführung*, Zeitschrift für Photogrammetrie und Fernerkundung, Vol. 63, No. 6, pp. 224–234.

Bähr H. P., T. Vögtle (eds) (1991) *Digitale Bildverarbeitung*, Wichmann, Karlsruhe.

Barr A., E. Feigenbaum (1982) *The Handbook of Artificial Intelligence*, Vol. 2, Kaufmann, Los Altos, CA.

Brachman R. (1977) *What's in a Concept: Structural Foundationns for Semantic Networks*, Bolt Beranek & Newman, Cambridge, Mass.

Bunke H. (1985) *Modellgesteuerte Bildanalyse*, B.G. Teubner, Stuttgart.

Dempster A. (1967) *Upper and lower probabilities, inferences based on a sample from a finite univariate population*, Biometrika, Vol. 55, pp. 515–528.

Dempster A. (1968) *A generalization of Bayesian inference*, Journal of the Royal Statistical Society, Series B, Vol. 30, pp. 205–247.

Findler N. V. (ed.) (1979) *Associative Networks*, Academic Press, Orlando.

Haralick R., L. Shapiro (1992a) *Computer and robot vision*, Vol. I, Addison-Wesley.

Haralick R., L. Shapiro (1992b) *Computer and robot vision*, Vol. II, Addison-Wesley.

Harbeck R. (ed.) (1994) *Das Geoinformationssystem ATKIS und seine Nutzung in Wirtschaft und Verwaltung*, Landesvermessungsamt Nordrhein–Westfalen, Bonn–Bad Godesberg.

Kummert F. (1992) *Flexible Steuerung eines sprachverstehenden Systems mit homogener Wissensbasis*, infix, Sankt Augustin.

Kummert F., H. Niemann, G. S. R. Prechtel (1993) *Control and explanation in a signal understanding environment*, Signal Processing, Vol. 3, pp. 111–145.

Lusti M. (1990) *Wissensbasierte Systeme*, B.I. Wissenschaftsverlag.

Niemann H., G. Sagerer, S. Schröder (1990a) *Ernest: A semantic network system for pattern understanding*, IEEE Transaction on Pattern Analysis and Machine Intelligence, pp. 257–269.

Niemann H., H. Brünig, R. Salzbrunn, S. Schröder (1990b) *A knowledge-based vision system for industrial applications*, Machine Vision and Applications, pp. 201–229.

Niemann H., H. Bunke (1987) *Künstliche Intelligenz in Bild- und Sprachanalyse*, B.G. Teubner, Stuttgart.

Paulus D. (1992) *Objektorientierte und wissensbasierte Bildverarbeitung*, Vieweg.

Quillian M. (1968) *Semantic Memory*, in M. Minsky (ed.), Semantic Information Processing, MIT Press, Cambridge, MA, pp. 227–270.

Quint F. (1995a) *Aerial image understanding using digital map based semantic models*, Euroconference GIS, Union Europeenne, Projet Capital Humain et Mobilité, DG XII, ENSG, St. Mande, France.

Quint F. (1995b) *An evidential merit function to guide search in a semantic network based image analysis system*, Technical Report IPF-FQ-11/95, University of Karlsruhe.

Quint F. (1995c) *Kartengestützte Bildauswertung*, in F. List (ed.), Vorträge 14. Wissenschaftlich-Technische Jahrestagung der DGPF, Vol. 3 of Publikationen der Deutschen Gesellschaft für Photogrammetrie und Fernerkundung, DGPF, Berlin, pp. 21–27.

Quint F., M. Sties (1995) *Map–based semantic modeling for the extraction of objects from aerial images*, in A. Grün, O. Kübler, P. Agouris (eds), Automatic Extraction of Man-Made Objects from Aerial and Space Images, Birkhäuser, Basel, pp. 307–316.

Quint F., M. Sties (1996a) *Evaluating model fidelity in an aerial image analysis system*, ISPRS, XVIIIth Congress, Intern. Arch. of Photogr. and Rem. Sens., Vol. 31, Part B3, pp. 669–674.

Quint F., M. Sties (1996b) *An evidential merit function to guide search in a semantic network based image analysis system*, in P. Perner, P. Wang, A. Rosenfeld (eds), Advances in Structural and Syntactical Pattern Recognition, Springer, Berlin, pp. 140–149.

Sagerer G., U. Ehrlich, F. Kummert, H. Niemann, E. Schukat-Talamazzini (1988) *A flexible control strategy with multilevel judgements for a knowledge based speech understanding system*, 9th International Conference on Pattern Recognition, Rome, pp. 788–790.

Schefe P. (1986) *Künstliche Intelligenz - Überblick und Grundlage*, Bibliographisches Institut.

Shafer G. (1976) *A mathematical theory of evidence*, Princeton Univerity Press.

Smets P. (1991) *The transferable belief model and other interpretations of Dempster-Shafer's model*, in P. Bonissone, M. Henrion, L. Kanal, J. Lemmer (eds), Uncertainty in artificial intelligence 6, North-Holland, Amsterdam, pp. 375–383.

Vögtle T., K.-J. Schilling (1995) *Wissensbasierte extraktion von siedlungsbereichen in der satellitenbildanalyse*, Zeitschrift für Photogrammetrie und Fernerkundung, Vol. 63, No. 5/95, pp. 199–207.

Winston P. (1987) *Künstliche Intelligenz*, Addison-Wesley.

Woods W. (1975) *What's in a link: Foundations for semantic networks*, in D. Bobrow, A. Collins (eds), Representation and Understanding, Academic Press, New York, pp. 35–82.

Semanitic Modeling for the Acquisition of Topografic Information from Images and Maps
ed. by W. Förstner and L. Plümer

Two Level Image Interpretation Based on Markov Random Fields

J. Klonowski, K.R. Koch

Institut für Theoretische Geodäsie
Rheinische Friedrich–Wilhelms–Universität Bonn
Nußallee 17, 53115 Bonn, Germany
Ph.: +49–228–732626; Fax: +49–228–733029
e–mail: koch@theor.geod.uni-bonn.de

Abstract

The interpretation of digital images is defined as a labeling process where the labels are represented by Markov random fields. Under the assumption that the objects to be identified in the image differ by their textures, pixels are labeled at the low level of the image analysis according to their affiliation to textures. Pixels with identical labels form the regions which are labeled at the high level of the image analysis to obtain the meaning of the objects. The textures at the low level are represented by the Gibbs distribution of the Markov random field for the gray values. Prior information on the labels concerning the textures is also introduced by a Gibbs distribution. The image model is therefore determined by these distributions. Information on regions and the relations between the regions is extracted from the image and represented by a Gibbs distribution. In addition prior knowledge on the objects to be interpreted is introduced by the same type of distribution. Hence, these distributions determine the object model. The densities of the prior information and of the data are combined by Bayes' theorem and the posterior density is maximized to estimate the labels. Posterior odds are computed for the labels to detect regions with possible misinterpretations. For these regions the segmentation needs to be improved. The results for the interpretation of two aerial multispectral photographs of urban areas are given.

1 Introduction

The interpretation of digital aerial images is a task which has to be solved automatically because of the ever increasing number of digital images taken by sensors in aeroplanes or satellites. There are no other economical methods than automatic interpretation to create or to update maps and geoinformation systems.

The interpretation of an image means to identify objects which are relevant to the content of the image. Objects have to be localized spatially and a meaning has to be attributed to them. Because of the complexity and of the ambiguous appearance of objects in an image, this task is restricted here to a two–dimensional approach. Two steps, or according to the level concept of image processing, a low and a high level will be used. Both levels build a homogeneous system because of the same underlying theory.

At the low level a digital image is segmented into picture primitives under the assumption that the objects contained in the image differ by their textures. The result is a partition into regions of same textures. At the high level a meaning is attributed to these regions by extracting characteristic data and by adding prior knowledge. The texture recognition at the level of the pixels as well as the attribution of a meaning to the regions are formulated as labeling processes. Additional knowledge, which is needed for the interpretation, is collected in the object model, which consists of geometrical, physical, structural and semantical parts. The appearance of the objects in the digital images is described in the image model.

For the image interpretation a method is needed which is based on uncertainty, since a digital image is noisy. By processing the image one gets random results at the low and the high level of the analysis. Random variables are therefore introduced for the labels of the pixels and the regions. They are represented by Markov random fields which are defined for graphs. The nodes coincide with the pixels or the regions and the paths indicate the neighbors. The modelling by Markov random fields therefore considers local dependencies of the random variables. The labels of the pixels and the labels of the regions depend on the labels of neighboring elements, pixels and regions are therefore seen in their local context.

Markov random fields have been used for a long time in digital image analysis. First autoregressive processes were applied and their Markov property proved (Woods 1972; Kashyap 1984; Chellappa 1985). Texture analysis (Kashyap and Chellappa 1983; Pan 1994) was the main field of application but also image restoration and smoothing were solved (Jain 1981). Markov random fields have Gibbs distributions according to the theorem of Hammersley and Clifford from 1971 which was proved for instance by Besag (1974) and Griffeath (1976). Geman and Geman (1984) were the first to apply this theorem to image processing. They also modelled edges by introducing line processes and introduced the now well–known Gibbs sampler.

Thus, Markov random fields are mostly used for low level image processing like restoration, edge detection and segmentation, see for instance (Koch and Schmidt 1994, p.299). But more and more applications appear at the high level processing, especially for image interpretation. Based on a segmented digital image Modestino and Zhang (1992) introduced a Markov random field for the labels of objects. Kristen and Munkelt (1992) extended their work by automizing the knowledge modelling and acquisation from sample images. High level interpretation and restoration using Markov random fields were investigated by Chou et al. (1993). Li (1994a,b) used a Bayesian formulation of the labeling problem with Markov random fields for matching features of a scene with different objects. The context dependencies were investigated and the relational description for objects were introduced by Köster (1995) using Markov random fields and Bayesian statistics. Indirect neighbors and cliques with three elements were used to improve the interpretation which is based on a segmented image. Kim and Yang (1996) start from a preliminary set of segmented regions and form region clusters together with the interpretation of the clusters.

In the following the image analysis is considered at two levels, at a low and at a high level. But both levels are designed as one system with the same underlying theory represented by Markov random fields. At the high level the method is based on the results of Köster (1995) but with the modification that a region which is not interpreted with a high probability gets the label null. The Gibbs distributions of the Markov random fields describe the texture and the prior information available for the textures. The image model is thus defined by these densities. Gibbs distributions are obtained by means of cliques which are based on neighborhoods so that relations between objects can be easily introduced. Prior information on the objects to be expected in the image is also introduced by Gibbs distributions, these distributions therefore determine the object model, too. The densities for the prior information and for the data are combined by Bayes' theorem to obtain the posterior densities for the labels. By maximizing these densities the values of the labels are estimated. Posterior odds of hypotheses for the labels at the high level of the image analysis indicate regions with week interpretations. These interpretations are improved by correcting the segmentations at the low level.

First ideas on this two level image interpretation have been outlined by Koch (1995). In the following the details together with results are represented. Section 2 deals with basic properties of Markov random fields and of Bayesian statistics and a modified Bayes' theorem is derived. The conditional densities for the texture recognition and for the interpretation are given in section 3 and 4. Section 5 describes the labeling process at the two levels of the image analysis. Results are shown in section 6 which are promising with respect to future applications.

2 Markov Random Field, Gibbs Distribution and Bayesian Analysis

As mentioned in the introduction, the interpretation of digital images is solved by means of Markov random fields. A random field $X(\boldsymbol{i})$

$$X(\boldsymbol{i}) = x_i \tag{2.1}$$

is defined as a sequence of random variables with values x_i. The elements $\boldsymbol{i}$ of the index set $\mathcal{I}$ are n-dimensional vectors. It is assumed, that $\mathcal{I}$ has a finite number I of elements, so that (2.1) is a discrete random field. If a random field

$$\boldsymbol{X}(\boldsymbol{i}) = [X_1(\boldsymbol{i}), \ldots, X_B(\boldsymbol{i})]' \tag{2.2}$$

consists of B components, it is called a vector random field. The index vector $\boldsymbol{i}$ is interpreted as a node of a graph.

A special class of random fields are the Markov random fields which fulfill the so-called Markov property. To define it, a neighborhood system has to be introduced for the graph with the index vectors $\boldsymbol{i}$ as nodes. Two nodes of the graph are neighbors, if a path joins them. All neighbors of node $\boldsymbol{i}$ form the neighborhood $\partial\boldsymbol{i}$ of that node. The Markov property can be formulated as a conditional density, see for instance (Koch and Schmidt 1994, p.180)

$$p(x_i|x_1,\ldots,x_{i-1},x_{i+1},\ldots,x_I) = p(x_i|\partial x_i) \; . \qquad (2.3)$$

Thus, the conditional density of the value x_i of a random variable $X(\boldsymbol{i})$ at node $\boldsymbol{i}$ depends only on the values in the neighborhood of node $\boldsymbol{i}$ abbreviated by ∂x_i. In a similar manner the Markov property of the vector random field (2.2) can be formulated, see for instance (Koch and Schmidt 1994, p.257).

The image interpretation is formulated as a labeling process at the low level and the high level of the image analysis. At the low level pixels are labeled according to their affiliation to textures. Let Ω be the set of pixels of a digital image

$$\Omega = \{r = (m,n)\;,\; 0 \le m \le M\;,\; 0 \le n \le N\}\;, \qquad (2.4)$$

where $r \in \Omega$ denotes a pixel at position (m,n) with $M+1$ being the total number of rows and $N+1$ the total number of columns of the image. The label ϵ_r, which is attributed to the pixel r, mainly depends on the labels of the neighborhood of the pixel r. Thus the labeling process is introduced as Markov random field $E(r)$

$$E(r) = \epsilon_r\;,\; r \in \Omega \qquad (2.5)$$

with

$$\epsilon_r = t\;,\; t \in \mathcal{T}\;,\; \mathcal{T} = \{1,\ldots,T\}\;, \qquad (2.6)$$

where t denotes the value of a label, that is the number of the texture to which the pixel r belongs, $\mathcal{T}$ is the set of values and T the total number of textures. The graph which underlies the Markov random field (2.5) is built by the regular grid of pixels. Clusters of pixels belonging to the same texture form the regions which give the picture primitives.

At the high level of the image analysis a label ϵ_p implying a meaning is attributed to a region. Again the label ϵ_p of a picture primitive p depends on the labels of the neighbors of the region p. Thus, a Markov random field is also introduced at the high level. Let $\mathcal{P}$ be the set of the regions with P elements

$$\mathcal{P} = \{1,\ldots,P\}\;, \qquad (2.7)$$

and $p \in \mathcal{P}$ an element of the set. In analogy to (2.5) we define the Markov random field $E(p)$

$$E(p) = \epsilon_p\;,\; p \in \mathcal{P} \qquad (2.8)$$

with

$$\epsilon_p = l \ , \ l \in \mathcal{L} \ , \ \mathcal{L} = \{0, 1, \ldots, L\} \ , \tag{2.9}$$

where ϵ_p denotes the label of the region p, l the value of the label and $L+1$ the total number of values of the set $\mathcal{L}$. A region with a week interpretation as determined by the posterior odds (5.6) of the hypothesis (2.29) obtains the label $l = 0$. Each picture primitive represents a node of the graph which underlies the random field (2.8). Thus, the graph at the high level is not based on a regular grid anymore.

According to the theorem of Hammerslay and Clifford Markov random fields have Gibbs distributions, see for instance (Koch and Schmidt 1994, p.261). Thus, if ϵ denotes the vector of a realization of the Markov random field (2.5) or (2.8)

$$\epsilon = (\epsilon_r) \quad \text{or} \quad \epsilon = (\epsilon_p) \ , \tag{2.10}$$

its joint density $p(\epsilon)$ is given by the Gibbs distribution

$$p(\epsilon) = \frac{1}{Z} \exp(-U(\epsilon)) \ , \tag{2.11}$$

where Z denotes the normalization constant, the so–called partition function, and $U(\epsilon)$ the energy of the random field. An alternative formulation used in connection with simulated annealing is

$$p(\epsilon) = \frac{1}{\bar{Z}} \exp(-U(\epsilon)/T_0) \ , \tag{2.12}$$

where T_0 denotes the temperature and $\bar{Z}$ the corresponding normalization constant. As mentioned before, the random fields are assumed to be discrete. Thus, the normalization constant Z in (2.11) is given by

$$Z = \sum_{\epsilon} \exp(-U(\epsilon)) \ . \tag{2.13}$$

The energy $U(\epsilon)$ in (2.11) is obtained by the sum over the potentials of so–called cliques. A clique c contains at least one node $\boldsymbol{i}$. In case of more than one node every node is the neighbor of every other node of that clique. The cliques c_q with q nodes will be collected in the set C_q

$$c_q \in C_q \quad \text{with} \quad q \in \{1, \ldots, Q\} \ . \tag{2.14}$$

The set $\mathcal{C}$ of all cliques of the graph is formed by the union of (2.14)

$$\mathcal{C} = C_1 \cup \ldots \cup C_Q \ . \tag{2.15}$$

Thus, the energy $U(\epsilon)$ of the Gibbs distribution (2.11) is defined by

$$U(\epsilon) = \sum_{c \in \mathcal{C}} U_c(\epsilon) \ , \tag{2.16}$$

where $U_c(\epsilon)$ denotes the potentials of the cliques of the graph for the random field. With (2.14) and (2.15) the energy follows with

$$U(\epsilon) = \sum_{c_1 \in C_1} U_{c_1}(\epsilon) + \ldots + \sum_{c_q \in C_q} U_{c_q}(\epsilon) + \ldots + \sum_{c_Q \in C_Q} U_{c_Q}(\epsilon) \,. \tag{2.17}$$

The conditional density (2.3) for the label ϵ_i at node $\boldsymbol{i}$ is obtained with (2.11) by

$$p(\epsilon_i|\partial\epsilon_i) \;\propto\; \exp\{-\sum_{c(\boldsymbol{i})\in C} U_c(\epsilon)\} \,, \tag{2.18}$$

see for instance (Koch and Schmidt 1994, p.262), where $\propto$ symbolizes the proportionality. The energy of the conditional density (2.18) includes only the potentials of the cliques $c(\boldsymbol{i})$ containing the node $\boldsymbol{i}$.

As mentioned, an automatic image interpretation is only feasible with additional knowledge about the objects of the image to be interpreted. This additional information can be processed by applying the very flexible Bayesian analysis. It allows not only parameter estimation but also hypothesis testing. Let the unknown labels be collected in the vector $\boldsymbol{\epsilon}$ and let $\boldsymbol{y}$ be the vector of observations. The posterior density $p(\boldsymbol{\epsilon}|\boldsymbol{y})$ of the unknown parameters $\boldsymbol{\epsilon}$ given the data $\boldsymbol{y}$ is obtained by Bayes' theorem, see for instance (Koch 1990, p.4)

$$p(\boldsymbol{\epsilon}|\boldsymbol{y}) = \frac{1}{C}\, p(\boldsymbol{\epsilon})p(\boldsymbol{y}|\boldsymbol{\epsilon}) \tag{2.19}$$

with

$$C = p(\boldsymbol{y}) = \sum_{\boldsymbol{\epsilon}} p(\boldsymbol{\epsilon})p(\boldsymbol{y}|\boldsymbol{\epsilon}) \tag{2.20}$$

in case of discrete random variables. Additional or prior information on the unknown parameters is introduced by the prior density $p(\boldsymbol{\epsilon})$. The density $p(\boldsymbol{y}|\boldsymbol{\epsilon})$ is called the likelihood function. According to (2.18) a conditional distribution will be used as prior density so that Bayes' theorem (2.19) has to be modified to accomodate conditional priors. This is readily accomplished by the definition of a conditional density

$$p(\boldsymbol{\epsilon}|\boldsymbol{y},\boldsymbol{x}) = \frac{p(\boldsymbol{\epsilon},\boldsymbol{y},\boldsymbol{x})}{p(\boldsymbol{y},\boldsymbol{x})} \,, \tag{2.21}$$

where $\boldsymbol{x}$ denotes a random vector which will be substituted by the appropriate vector in the following chapters. By applying the definition (2.21) again to express $p(\boldsymbol{\epsilon},\boldsymbol{y},\boldsymbol{x})$ and $p(\boldsymbol{y},\boldsymbol{x})$ one obtains

$$p(\boldsymbol{\epsilon}|\boldsymbol{y},\boldsymbol{x}) \;\propto\; p(\boldsymbol{\epsilon}|\boldsymbol{x})p(\boldsymbol{y}|\boldsymbol{\epsilon},\boldsymbol{x}) \,, \tag{2.22}$$

where $\boldsymbol{y}$ and $\boldsymbol{x}$ are assumed as given and the prior information is now introduced by the conditional density $p(\boldsymbol{\epsilon}|\boldsymbol{x})$.

The unknown labels $\boldsymbol{\epsilon}$ are estimated by maximizing the posterior density

$$\bar{\boldsymbol{\epsilon}} = \arg\max_{\boldsymbol{\epsilon}} \; p(\boldsymbol{\epsilon}|\boldsymbol{y},\boldsymbol{x}) \,, \tag{2.23}$$

where $\overline{\epsilon}$ is called the MAP estimate of ϵ. It is computed either by a deterministic or a stochastic method. The deterministic approach is based on the pseudo–likelihood method by assuming that the joint posterior density $p(\epsilon|\boldsymbol{y},\boldsymbol{x})$ is obtained by the product of the conditional densities for every node of the random field $E(r)$ or $E(p)$

$$p(\epsilon|\boldsymbol{y},\boldsymbol{x}) = \prod_i p(\epsilon_i|\boldsymbol{y},\boldsymbol{x}) \,. \tag{2.24}$$

This leads to the ICM (iterated conditional modes) procedure (Besag 1986), where the MAP estimate is iteratively applied to the conditional density of every node

$$\overline{\epsilon}_i = \arg\max_{\epsilon_i} \; p(\epsilon_i|\boldsymbol{y},\boldsymbol{x}) \,. \tag{2.25}$$

The stochastic method is based on simulated annealing (Kirkpatrick *et al.* 1983). It maximizes according to (2.23) the joint density $p(\epsilon|\boldsymbol{y},\boldsymbol{x})$ by actually evaluating the conditional density $p(\epsilon_i|\boldsymbol{y},\boldsymbol{x})$ in (2.25). A variant of this method was introduced by Geman and Geman (1984) using the Gibbs sampler. The ICM method is very fast, it is applied for the very time consuming labeling of the pixels at the low level of the image analysis. The image interpretation at the high level takes less time, but the results of the ICM method and the simulated annealing agree very well, as was shown by Köster (1995), so that the ICM method is applied only.

As already mentioned, the posterior density for the labels also allows the testing of hypotheses. Let $\mathcal{E}$ be the parameter space spanned by the vector ϵ of the unknown labels. Let $\mathcal{E}_0$ and $\mathcal{E}_1$ be two disjoint subsets of $\mathcal{E}$, hence $\mathcal{E}_0 \cap \mathcal{E}_1 = \emptyset$, then the null hypothesis H_0 and the alternative hypothesis H_1 are defined by

$$H_0 : \epsilon \in \mathcal{E}_0 \quad \text{versus} \quad H_1 : \epsilon \in \mathcal{E}_1 \,. \tag{2.26}$$

The probability $P(H_j|\boldsymbol{y},\boldsymbol{x})$ of the hypothesis H_j with $j \in \{0,1\}$ given the observation vector $\boldsymbol{y}$ and the conditioning $\boldsymbol{x}$ is computed by the posterior density

$$P(H_j|\boldsymbol{y},\boldsymbol{x}) = P(\epsilon \in \mathcal{E}_j|\boldsymbol{y},\boldsymbol{x}) = \sum_{\epsilon\in\mathcal{E}_j} p(\epsilon|\boldsymbol{y},\boldsymbol{x}) \,. \tag{2.27}$$

To decide whether to accept or reject the null hypothesis the posterior odds are computed, see for instance (Koch 1990, p.43)

$$\frac{P(H_0|\boldsymbol{y},\boldsymbol{x})}{P(H_1|\boldsymbol{y},\boldsymbol{x})} \,. \tag{2.28}$$

If this ratio is larger than 1 the null hypothesis H_0 is accepted, otherwise it is rejected. In the following the hypothesis for the label ϵ_p in (2.9)

$$H_0 : \epsilon_p = l \quad \text{versus} \quad H_1 : \epsilon_p = k \quad \text{with} \quad l \in \mathcal{L} \quad \text{and} \quad k \in \mathcal{L} \setminus l \tag{2.29}$$

is tested. Posterior odds close to one for this hypothesis indicate that the labeling of the region might not be correct and needs to be improved. Such a region obtains the label $\epsilon_p = 0$.

3 Conditional Density for Low Level Image Processing

At the low level of the image analysis the pixels are labeled according to their affiliation to textures by applying the Markov random field $E(r)$ in (2.5). Clusters of pixels rather than a few pixels are generally attributed to one texture. This fact is used as prior information. Its Gibbs distribution $p(\epsilon_r|\partial\epsilon_r)$ is introduced as a special case of the so-called auto model (Besag 1974), which considers only cliques with one and two nodes

$$p(\epsilon_r|\partial\epsilon_r) \propto \exp\Big\{-[\alpha_t + \sum_{e\in\partial r_\epsilon} \beta_e(I(\epsilon_r, \epsilon_{r+e}) + I(\epsilon_r, \epsilon_{r-e}))]\Big\}, \tag{3.1}$$

with the indicator function

$$I(\epsilon_{r_1}, \epsilon_{r_2}) = \begin{cases} 1, & \text{if} \quad \epsilon_{r_1} \neq \epsilon_{r_2} \\ 0, & \text{if} \quad \epsilon_{r_1} = \epsilon_{r_2}. \end{cases} \tag{3.2}$$

The index e denotes a neighbor of the pixel r in its neighborhood ∂r_ϵ, α_t and β_e the parameters of the density. The parameter α_t controls the number of pixels belonging to one texture $t \in \mathcal{T}$. Large values of α_t with $\alpha_t > 0$ cause small density values, thus fewer pixels for the texture t. The parameters β_e with $\beta_e > 0$ govern the affiliation of pixels to the same texture and also the orientation of the texture (Koch and Schmidt 1994, p.313).

Information on the textures is coming from the measurements of the gray levels of the digital image. These measurements are given for different spectral bands. Thus, the appropriate random field is a vector random field with Markov property, since the gray value of a pixel r in the channel b is mainly influenced by the gray values of the neighborhood of r in the same spectral band b. A component $Y(r)_b$ of the random field of the measurements in channel b at the pixel r is expressed by

$$Y(r)_b = y_{rb} \tag{3.3}$$

with

$$b \in \{1, \ldots, B\}, \quad 0 \leq y_{rb} \leq 255 \tag{3.4}$$

and B being the number of channels, y_{rb} being the measurement. Let y_{rb} be collected in the vector $\boldsymbol{y}_r$ of measurements at the pixel r

$$\boldsymbol{y}_r = [y_{r1}, \ldots, y_{rB}]'. \tag{3.5}$$

By assuming independent and normally distributed measurements y_{rb} the likelihood function is obtained as a special case of a Gibbs distribution

$$p(\boldsymbol{y}_r|\partial\boldsymbol{y}_r, \epsilon_r, \partial\epsilon_r) \propto \exp\Big\{-\sum_{b=1}^{B}\{\frac{1}{2\sigma_{tb}^2}[y_{rb} - \mu_{tb} - \sum_{e\in\partial r_y} \beta_{etb}(y_{r+e,b} - \mu_{tb} + y_{r-e,b} - \mu_{tb})]^2\}\Big\}. \tag{3.6}$$

The parameters of the density are the variance σ_{tb}^2 of an observation y_{rb}, the mean gray value μ_{tb} of the texture t and the parameters β_{etb}, which describe the texture t at the channel b with $\epsilon_r = \partial\epsilon_r = t$. The index e in (3.6) again denotes a neighbor of the pixel r in its neighborhood ∂r_y.

It should be mentioned that in the prior distribution (3.1) as well as in the likelihood function (3.6) a local spatial context is considered. With (3.1) and (3.6) the conditional posterior density for the label ϵ_r of pixel r is obtained by applying the modified Bayes' theorem (2.22)

$$p(\epsilon_r|\boldsymbol{y}_r, \partial\boldsymbol{y}_r, \partial\epsilon_r) \propto p(\epsilon_r|\partial\epsilon_r)p(\boldsymbol{y}_r|\partial\boldsymbol{y}_r, \epsilon_r, \partial\epsilon_r) , \qquad (3.7)$$

since the prior density (3.1) can be assumed as being conditionally dependent on $\partial\boldsymbol{y}_r$. The unknown labels ϵ_r are estimated by applying (2.23).

4 Conditional Density for Image Interpretation

At the high level of the image processing the regions p obtained by the segmentation have to be labeled by using the Markov random field $E(p)$ defined in (2.8) with the realization $\epsilon = (\epsilon_p)$ and $p \in \mathcal{P}$.

Prior information on the label ϵ_p follows from the frequencies of the occurrence of objects and neighboring objects. This knowledge is derived from the image which needs to be interpreted or from similar images. It is part of the object model. The prior density of the label ϵ_p is obtained with (2.17) from (2.18)

$$p(\epsilon_p|\partial\epsilon_p) \propto \exp\Big\{- U_{p11}(\epsilon) - \ldots - \frac{1}{S_{pq}}\sum_{s=1}^{S_{pq}} U_{pqs}(\epsilon) - \ldots - \frac{1}{S_{pQ}}\sum_{s=1}^{S_{pQ}} U_{pQs}(\epsilon)\Big\}$$
$$\text{with} \quad q \in \{1, \ldots, Q\}\,, \; s \in \{1, \ldots, S_{pq}\}\,. \qquad (4.1)$$

S_{pq} denotes the number of cliques with q nodes. It varies with the regions p so that it is introduced as a normalization constant of the clique potentials $U_{pqs}(\epsilon)$, which is very helpful for the numerical analysis. However, it violates the definition of the energy (2.17) where the potentials of the cliques have unique values. Nevertheless, it can be accepted since the ICM method (2.25) only is applied for the MAP estimate.

Prior information may be also expressed by the frequencies mentioned above so that the conditional density $p(\epsilon_p|\partial\epsilon_p)$ can be written as the product of the densities $p_{pq}(\epsilon)$ of the cliques with q nodes

$$p(\epsilon_p|\partial\epsilon_p) = p_{p1}(\epsilon)\ldots p_{pq}(\epsilon)\ldots p_{pQ}(\epsilon) = \prod_{q=1}^{Q} p_{pq}(\epsilon) \qquad (4.2)$$

$$= \prod_{q=1}^{Q}\Big(\prod_{s=1}^{S_{pq}} p_{pqs}(\epsilon)\Big)^{1/S_{pq}} , \qquad (4.3)$$

where $p_{pqs}(\epsilon)$ denotes the contribution of a clique within the cliques of q sites. The number S_{pq} has been used again for the normalization. By comparing the right hand sides of (4.1) and (4.3) we conclude

$$(p_{pqs}(\epsilon))^{1/S_{pq}} \propto \exp\Big\{-\frac{1}{S_{pq}}U_{pqs}(\epsilon)\Big\} \tag{4.4}$$

and obtain the relationship between the frequency $p_{pqs}(\epsilon)$ and the potential $U_{pqs}(\epsilon)$ of the cliques

$$p_{pqs}(\epsilon) \propto \exp\Big\{-U_{pqs}(\epsilon)\Big\}\,. \tag{4.5}$$

The density $p_{p11}(\epsilon)$ for the one–site cliques is determined by the frequencies of the occurrence of the objects as mentioned above. In case of $l = 0$ a small value is chosen for $p_{p11}(\epsilon)$, since the estimation of a label $l = 0$ is discouraged. For cliques with more than one site $p_{pqs}(\epsilon) = 1$ is chosen, if the neighborhood determined by the labels exists in the image which is analysed. This information is contained in the object model. A small value for $p_{pqs}(\epsilon)$ is chosen, if the neighborhood is unlikely or if one site has the label $l = 0$.

Data which characterize the regions and relations between regions are extracted from the image. These observations should be invariant to geometrical and photometrical transformations of the regions and their gray values, the observations shown in Tabular 1 were therefore chosen. Due to the representation (2.17) of the energy of their Gibbs distribution, the observations are therefore extracted for cliques. The maximum number of nodes chosen here is three.

One–site cliques		
Form	$Fo(p)$	$= \frac{\lambda_1}{\lambda_2}$ for $\lambda_2 \neq 0$
Roundness	$Ro(p)$	$= \frac{D^2(p)}{4\pi F(p)}$ for $F(p) \neq 0$
Compactness	$Co(p)$	$= \frac{F(p)}{F_S(p)}$ for $F_S(p) \neq 0$
Hu–Invariants	Hu_1	$= \frac{\mu_{20}+\mu_{02}}{F(p)^2}$ for $F(p) \neq 0$
	Hu_2	$= \frac{(\mu_{20}-\mu_{02})^2+4\mu_{11}^2}{F(p)^4}$ for $F(p) \neq 0$
Two–sites cliques		
Ratio of Areas	RA_{po}	$= \frac{F(p)}{F(o)}$ for $F(o) \neq 0$
Difference of Orientations	DO	$= \lvert\phi(p) - \phi(o)\rvert$
Ratio of Mean Gray Values	$RG_{po,b}$	$= \frac{G_b(p)}{G_b(o)}$ for $G_b(o) \neq 0$
Ratio of Variances	$RV_{po,b}$	$= \frac{\sigma^2_{G_b}(p)}{\sigma^2_{G_b}(o)}$ for $\sigma^2_{G_b}(o) \neq 0$
Three–sites cliques		
Sum of Ratios of Areas	RA_{pow}	$= RA_{po} + RA_{pw} + RA_{ow}$
Sum of Differences of Orientations	DO_{pow}	$= DO_{po} + DO_{pw} + DO_{ow}$
Sum of Ratios of Mean Gray Values	$RG_{pow,b}$	$= RG_{po,b} + RG_{pw,b} + RG_{ow,b}$
Sum of Ratios of Variances	$RV_{pow,b}$	$= RV_{po,b} + RV_{pw,b} + RV_{ow,b}$

Tab. 1: Observations for the cliques

Observations for one–site cliques describe the appearence of a region. The following quantities are used:

- λ_1, λ_2 eigenvalues of the matrix of the second central moments,
- $D(p)$ perimeter and $F(p)$ area of the region p,
- $F_S(p)$ area of the region enclosing rectangle and
- $\mu_{20}, \mu_{11}, \mu_{02}$ second central moments.

The observations for two–sites cliques introduce relations between two regions. The ratio of the areas expresses a relation of the largeness of the regions. The difference of the angles $\phi(p)$ describes the orientation of regions, especially parallelity and orthogonality. A relation of the brightness is given by the ratio of the mean gray values $G(p)$. The variance $\sigma^2_{G_b}(p)$ of the gray values is a measure for the homogeneity of a region p, the ratio for two regions is also introduced as an observation. Since the image is multispectral, there is a ratio of gray values and variances in each channel b.

A three–sites clique is composed of three two–sites cliques. Therefore observations are obtained by adding the ratios and differences of the two–sites cliques as shown in Tabular 1. The introduction of three–sites cliques enlarges the context which is recognized, since an extra path with new information is used for the interpretation.

These observations are introduced for each region p and build the vector $\boldsymbol{y}_p$ of observations with the components

$$\boldsymbol{y}_p = (y_{pqsv}) \quad \text{with} \quad q = \{1, \dots, Q\},\ s = \{1, \dots, S_{pq}\},\ v = \{1, \dots, V_q\}\,. \tag{4.6}$$

For each clique with q sites V_q observations are available. It is assumed that the components of $\boldsymbol{y}_p$ are normally distributed and independent despite of the fact that observations for two–sites cliques were added. Thus, the likelihood function is obtained again as a special Gibbs distribution by

$$\begin{aligned} p(\boldsymbol{y}_p|\partial\boldsymbol{y}_p, \epsilon_p, \partial\epsilon_p) \propto \exp\Big\{ & -\frac{1}{V_1}\sum_{v=1}^{V_1}\frac{1}{2\sigma^2_{11v\epsilon}}(y_{p11v} - \mu_{11v\epsilon})^2 - \cdots \\ & - \frac{1}{S_{pq}V_q}\sum_{s=1}^{S_{pq}}\sum_{v=1}^{V_q}\frac{1}{2\sigma^2_{qsv\epsilon}}(y_{pqsv} - \mu_{qsv\epsilon})^2 - \cdots \\ & - \frac{1}{S_{pQ}V_Q}\sum_{s=1}^{S_{pQ}}\sum_{v=1}^{V_Q}\frac{1}{2\sigma^2_{Qsv\epsilon}}(y_{pQsv} - \mu_{Qsv\epsilon})^2\Big\}\,. \end{aligned} \tag{4.7}$$

The parameters $\mu_{qsv\epsilon}$ und $\sigma^2_{qsv\epsilon}$ of the density, which are the expected value and the variance of the observation y_{pqsv}, depend on the labels ϵ_p and $\partial\epsilon_p$. These parameters are given by the object model.

The conditional posterior density of the label ϵ_p is obtained in analogy to (3.7) from the modified Bayes' theorem (2.22)

$$p(\epsilon_p|\boldsymbol{y}_p, \partial\boldsymbol{y}_p, \partial\epsilon_p) \propto p(\epsilon_p|\partial\epsilon_p)p(\boldsymbol{y}_p|\partial\boldsymbol{y}_p, \epsilon_p, \partial\epsilon_p)\,. \tag{4.8}$$

5 The Labeling Process

For the low level of the image analysis the conditional posterior density of the label ϵ_r at the pixel r follows from (3.7) with (3.1) and (3.6) by

$$p(\epsilon_r|\boldsymbol{y}_r, \partial\boldsymbol{y}_r, \partial\epsilon_r) \propto \exp\Big\{-[\alpha_t + \sum_{e\in\partial r_\epsilon} \beta_e(I(\epsilon_r, \epsilon_{r+e}) + I(\epsilon_r, \epsilon_{r-e}))]$$
$$-\sum_{b=1}^{B}\{\frac{1}{2\sigma_{tb}^2}[y_{rb} - \mu_{tb} - \sum_{e\in\partial r_y} \beta_{etb}(y_{r+e,b} - \mu_{tb} + y_{r-e,b} - \mu_{tb})]^2\}\Big\}\,. \quad (5.1)$$

The parameters $\alpha_t, \beta_e, \sigma_{tb}^2, \mu_{tb}$ and β_{etb} in (5.1) are unknown and have to be estimated. By inserting these estimates in (5.1), the labels ϵ_r are then determined by (2.23).

The unknown parameters σ_{tb}^2, μ_{tb} and β_{etb} in (3.6) and (5.1) depend on the textures. For their estimation it is assumed that training sets are available for each texture $t \in \mathcal{T}$. These training sets are part of the object model. Given the gray values y_{rb} of the training set for the texture t at the channel b, the conditional density of σ_{tb}^2, μ_{tb} and β_{etb} follows from (3.6) by

$$p(\sigma_{tb}^2, \mu_{tb}, \beta_{etb}|y_{rb}, \partial y_{rb}, \epsilon_r = \partial\epsilon_r = t)$$
$$\propto \exp\Big\{-\frac{1}{2\sigma_{tb}^2}[y_{rb} - \mu_{tb} - \sum_{e\in\partial r_y} \beta_{etb}(y_{r+e,b} - \mu_{tb} + y_{r-e,b} - \mu_{tb})]^2\Big\}\,. \quad (5.2)$$

This density is defined for each pixel $r \in \Omega_{tb}$ belonging to the set Ω_{tb} of pixels in the channel b and the texture t. By applying the pseudo–likelihood method (2.24) the joint density of σ_{tb}^2, μ_{tb} and β_{etb} is obtained by summing over $r \in \Omega_{tb}$. Thus, the exponent of the joint density is given by

$$-\frac{1}{2\sigma_{tb}^2}\sum_{r\in\Omega_{tb}}[y_{rb} - \mu_{tb} - \sum_{e\in\partial r_y} \beta_{etb}(y_{r+e,b} - \mu_{tb} + y_{r-e,b} - \mu_{tb})]^2\,. \quad (5.3)$$

By applying the MAP estimate of σ_{tb}^2 and β_{etb} this sum has to be minimized leading to the well known method of least squares. The parameter μ_{tb} is estimated simply by forming the mean.

The unknown parameters α_t and β_e in (3.1) and (5.1) depend on the segmentation. To estimate them, the segmentation has to be known. Given a segmentation, the conditional density of α_t and β_e is obtained from (3.1) by

$$p(\alpha_t, \beta_e|\epsilon_r, \partial\epsilon_r) \propto \exp\Big\{-[\alpha_t + \sum_{e\in\partial r_\epsilon} \beta_e(I(\epsilon_r, \epsilon_{r+e}) + I(\epsilon_r, \epsilon_{r-e}))]\Big\}\,. \quad (5.4)$$

This density is defined for each pixel $r \in \Omega$. The joint density again is obtained by the pseudo–likelihood method (2.24). MAP estimates of α_t and β_e are determined from (2.23) by searching for the maximum of the density for discrete values of α_t and β_e.

As mentioned, the segmentation has to be known to estimate α_t and β_e. However, the goal of (5.1) is to obtain a segmented image. The labeling according to (5.1) and the estimation of α_t and β_e has therefore to be solved iteratively. Starting with the estimates of σ_{tb}^2, μ_{tb} and β_{etb} only the part in (5.1) stemming from the likelihood function (3.6) is applied to estimate the labels ϵ_r. The resulting segmentation is used to estimate α_t and β_e. Then the complete density (5.1) is applied for a second segmentation, which is used again to estimate α_t and β_e. The iterations stop if less than 0.23% of the pixels get new labels. This state is reached after about 8 iterations. As mentioned the ICM method is applied to estimate the labels. The results show that especially close to the boundaries of textures some isolated pixels obtain incorrect labels. These misclassifications are removed by eliminating regions containing less than 30 pixels.

Neighboring pixels with the same label are collected to form a region. At the high level of the image analysis a meaning is attributed to these regions. Therefore, the conditional posterior density (4.8) of the label ϵ_p resulting from (4.1) and (4.7) has to be maximized

$$p(\epsilon_p|\boldsymbol{y}_p, \partial\boldsymbol{y}_p, \partial\epsilon_p) \propto \exp\Big\{-\sum_{q=1}^{Q}\frac{1}{S_{pq}}\sum_{s=1}^{S_{pq}}[U_{pqs}(\epsilon)+\frac{1}{V_q}\sum_{v=1}^{V_q}\frac{1}{2\sigma_{qsv\epsilon}^2}(y_{pqsv}-\mu_{qsv\epsilon})^2]\Big\}. \quad (5.5)$$

The MAP estimates of the labels ϵ_p are determined by the ICM method based on (2.24). The ICM method depends on good approximate values for the labels ϵ_p. They are obtained by the segmentation, which already identifies objects. With these approximations the ICM method converges in about 4 iterations to the state that less than 6% of the labels of the regions change.

Image interpretation with Markov random fields provides an estimated label $\bar{\epsilon}_p = l$ for a region p with maximum probability and in addition the density values of the labels $k \in \mathcal{L}\setminus l$. To evaluate the quality of the labeling, the posterior odds (2.28) for the estimated label and the alternative labels are computed by the posterior density (5.5). We set $\bar{\epsilon}_p = 0$, if

$$\frac{p(\epsilon_p = l|\boldsymbol{y}_p, \partial\boldsymbol{y}_p, \partial\epsilon_p)}{p(\epsilon_p = k|\boldsymbol{y}_p, \partial\boldsymbol{y}_p, \partial\epsilon_p)} < W \quad \text{with} \quad l \in \mathcal{L}\,,\ k \in \mathcal{L}\setminus l\,, \quad (5.6)$$

and for instance $W = 19$. This value results from assuming $p(\epsilon_p = l|\boldsymbol{y}_p, \partial\boldsymbol{y}_p, \partial\epsilon_p) = 0.95$ and $p(\epsilon_p = k|\boldsymbol{y}_p, \partial\boldsymbol{y}_p, \partial\epsilon_p) = 0.05$.

6 Results

The interpretation presented in section 5 has been applied to a part of the aerial photograph "glandorf", which constitutes a test data set of the ISPRS Working Group III/3 (Fritsch *et al.* 1994). The image consists of the three channels red, green and blue in the RGB color model. Since the image contains many shadows, which appear in each of the channels, the gray values of the channels were transformed into the HSI color model (hue, saturation and intensity). Hue is a color attribute that describes a pure color, whereas saturation gives a measure of the degree to which a pure color is diluted by white light. The intensity component, which is the mean of the R, G and B values at one pixel, is totally decoupled from the color information in the image. Thus, in the HSI color model the shadows only appear in the intensity component. The transformation is derived for instance by Gonzales and Woods (1992, p.229). Figure 1 as well as Figure 4 show the intensity by means of gray values of the part of image "glandorf".

For the labeling process at the low level eight training sets were selected from the digital image for the segmentation. They represent streets, houses with olive roofs, houses with red roofs, lawns, meadows, bushes, townhouses and shadows from trees or houses. Figure 2 shows the result of the segmentation. It contains 318 regions, since many regions are very small. Based on this segmentation nine objects had to be identified: streets (s), houses with olive roofs (ho), houses with red roofs (hr), annexes or garages (ga), lawns (l), meadows (m), bushes (b), green areas within streets (g), townhouses (t). In order to interpret with a high probability, $W = 19$ was chosen in (5.6). The results are given in Figure 3, where the regions with $l = 0$ and the identified regions are represented by gray values. Uninterpreted regions with the label $l = 0$ cover 10.2% of the area of the image. In addition the townhouses were misinterpreted and also some houses.

Three additional training sets representing houses and townhouses with roofs of different colors were therefore introduced leading to the segmentation of Figure 5, which contains 357 regions. The interpretation of these regions with $W = 19$ in (5.6) is shown in Figure 6. The area covered by regions with label $l = 0$ dropped to 7.6%. However, some regions were still incorrectly interpreted for instance the region in the upper left corner which is identified as a street. From a visual inspection one can judge that these misclassified objects cover an area of about 14% so that regions incorrectly identified and regions with label $l = 0$ together cover an area of about 22%.

The interpretation was also applied to a part of an aerial photograph consisting of the three channels red, green and blue, where the red color is close to infrared. The grey values of the green channel are shown in Figure 7. After trying different training sets together with the interpretations seven training sets were finally chosen for the labeling process at the low level: streets, houses with dark roofs, houses with light roofs, lawns, bushes and two different agricultural fields. The results of the segmentation are shown in Figure 8.

Based on this segmentation five objects had to be identified: streets (s), houses (h), lawns (l), bushes (b), fields (f). $W = 9$ was chosen in (5.6). The results of the interpretation are given in Figure 9, where again the regions with the label $l = 0$ and the identified regions are shown by gray values. Regions with $l = 0$ cover 4.2% of the area. A visual inspection of the interpretation shows that the misclassified

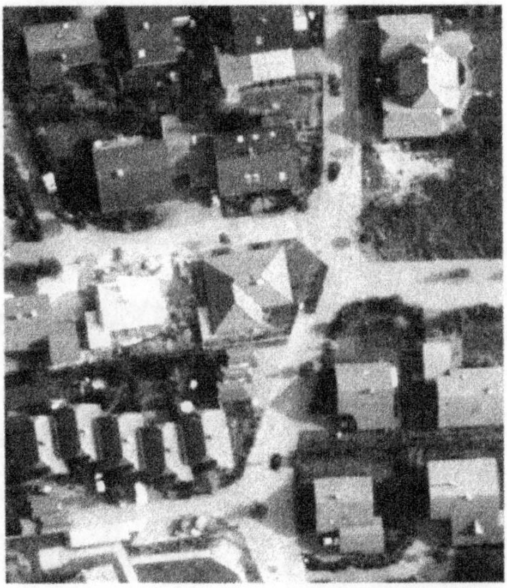

Fig. 1: Intensity of colors of the digital image

Fig. 2: Segmented HSI image based on eight training sets

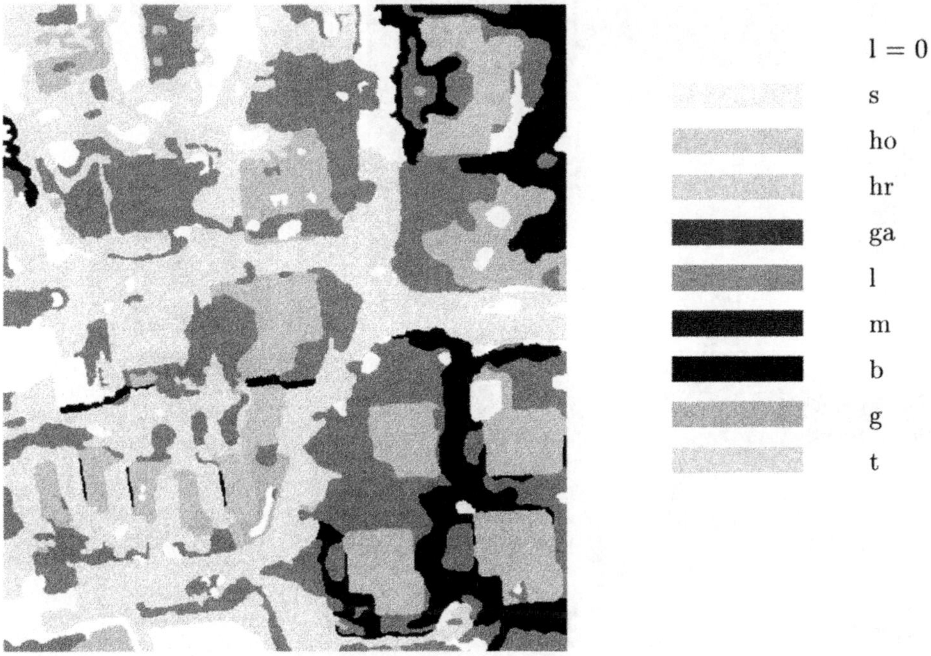

Fig. 3: Interpretation based on eight training sets

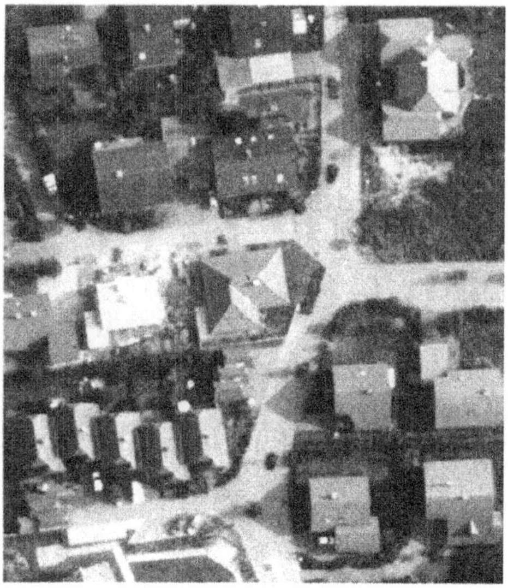

Fig. 4: Intensity of colors of the digital image

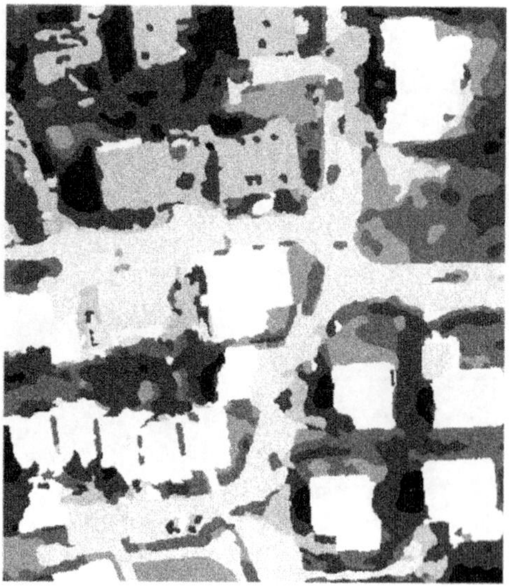

Fig. 5: Segmented HSI image based on eleven training sets

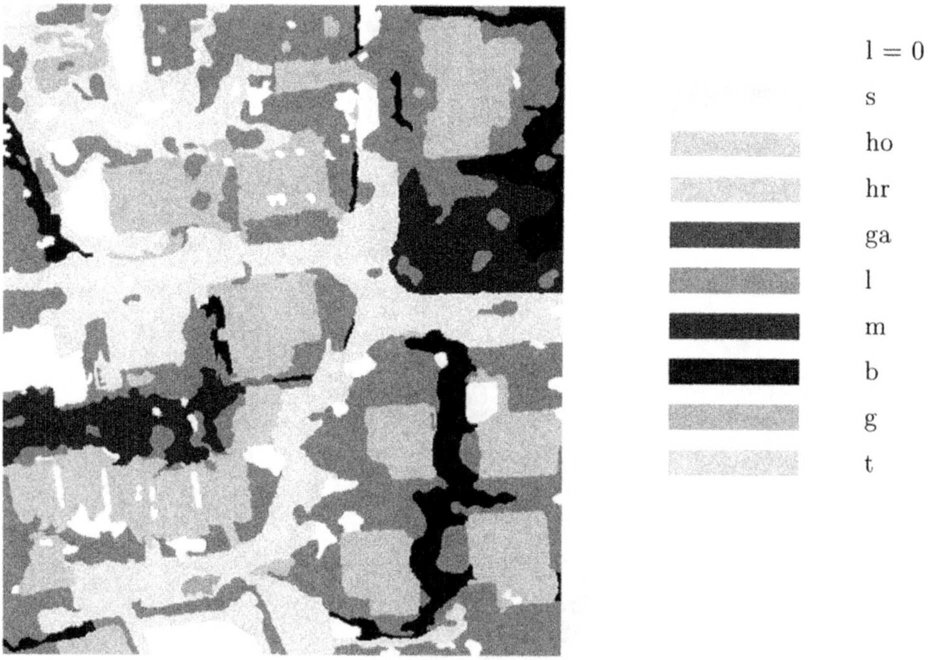

Fig. 6: Interpretation based on eleven training sets

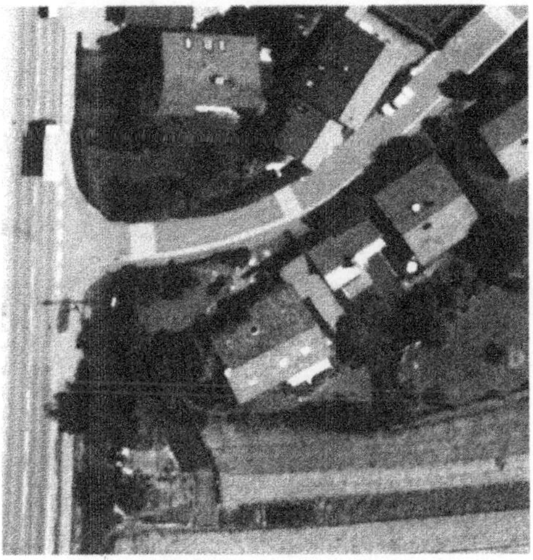

Fig. 7: Grey values of the green channel

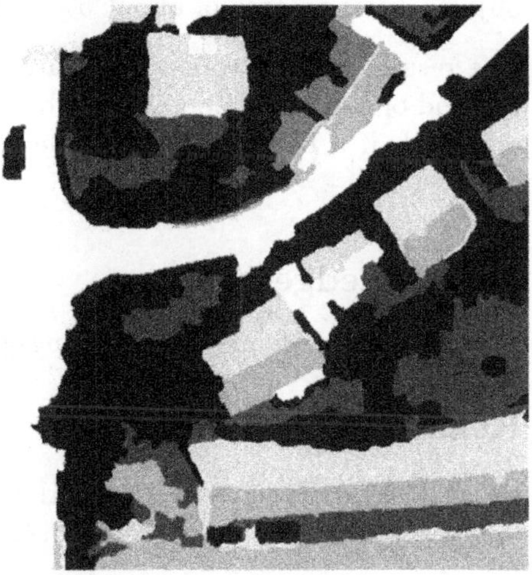

Fig. 8: Segmented image

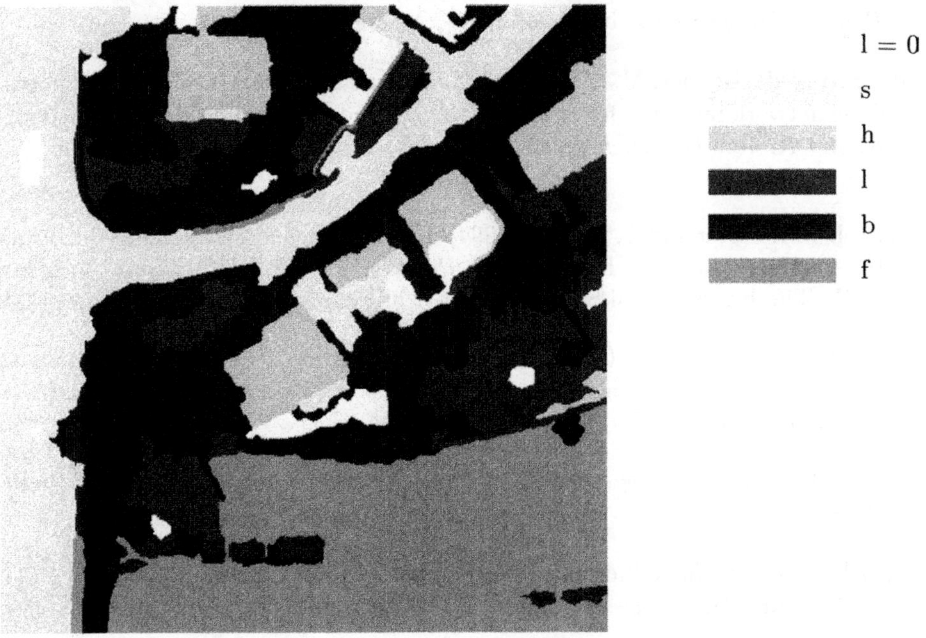

Fig. 9: Interpretation

objects cover an area of about 5% or 9% together with the regions with $l = 0$.

The results of the two examples are encouraging for a further development of this method of interpretation. A considerable improvement of the interpretation can be expected, if a digital terrain model becomes available for the digital image to be analysed.

Acknowledgements

The support of Deutsche Forschungsgemeinschaft (DFG) is gratefully mentioned.

References

Besag, J.E. (1974) Spatial interaction and the statistical analysis of lattice systems. *J Royal Statist Society*, B 36:192–236.

Besag, J.E. (1986) On the statistical analysis of dirty pictures. *J Royal Statist Society*, B 48:259–302.

Chellappa, R. (1985) Two–dimensional discrete Gaussian Markov random field models for image processing. In: Kanal, L.N. and A. Rosenfeld (Eds.), *Progress in Pattern Recognition 2.* North–Holland, Amsterdam, 79–112.

Chou, P.B., P.R. Cooper, M.J. Swain, C.M. Brown and L.E. Wixson (1993) Probabilistic network inference for cooperative high and low level vision. In: Chellappa, R. and A. Jain (Eds.), *Markov Random Fields.* Academic Press, Boston, 211–243.

Fritsch, D., M. Sester and T. Schenk (1994) Test on image understanding. In: Ebner, H., C. Heipke and K. Eder (Eds.), *Proc ISPRS Comm III Symp, München.* SPIE–The International Society for Optical Engineering, Bellingham, 243–248.

Geman, S. and D. Geman (1984) Stochastic relaxation, Gibbs distributions, and the Bayesian restoration of images. *IEEE Trans Pattern Anal Machine Intell*, PAMI–6:721–741.

Gonzales, R.C. and R.E. Woods (1992) *Digital Image Processing.* Addison–Wesley, Reading.

Griffeath, D. (1976) Introduction to random fields. In: Kemeny, J.G., J.L. Snell and A.M. Knapp (Eds.), *Denumerable Markov Chains.* Springer, Berlin, 425–458.

Jain, A.K. (1981) Advances in mathematical models for image processing. *Proc IEEE*, 69:502–528.

Kashyap, R.L. (1984) Characterization and estimation of two–dimensional ARMA models. *IEEE Trans Inform Theory*, IT–30:736–745.

Kashyap, R.L. and R. Chellappa (1983) Estimation and choice of neighbors in spatial-interaction models of images. *IEEE Trans Inform Theory*, IT–29:60–72.

Kim, I.Y. and H.S. Yang (1996) An integration scheme for image segmentation and labeling based on Markov random field model. *IEEE Trans Pattern Anal Machine Intell*, 18:69–73.

Kirkpatrick, S., C.D. Gelatt and M.P. Vecchi (1983) Optimization by simulated annealing. *Science*, 220:671–680.

Koch, K.R. (1990) *Bayesian Inference with Geodetic Applications*. Springer, Berlin.

Koch, K.R. (1995) Markov random fields for image interpretation. *Z Photogrammetrie und Fernerkundung*, 63:84–90, 147.

Koch, K.R. and M. Schmidt (1994) *Deterministische und stochastische Signale*. Dümmler, Bonn.

Köster, M. (1995) *Kontextsensitive Bildinterpretation mit Markoff–Zufallsfeldern*. Reihe C, 444. Deutsche Geodätische Kommission, München.

Kristen, H. and O. Munkelt (1992) Markov–Feld–basierte Bildinterpretation mit automatisch generierten Datenbasen. In: Fuchs, S. and R. Hoffmann (Eds.), *Mustererkennung 1992*. Springer, Berlin, 50–57.

Li, S.Z. (1994a) A Markov random field model for object matching under contextual constraints. *Proc Conf Computer Vision Pattern Recognition, Seattle*, IEEE Computer Society Press, Los Alamitos, 866–869.

Li, S.Z. (1994b) Markov random field models in computer vision. In: Eklundh, J.-O. (Ed.), *Computer Vision – ECCV'94*. Springer, Berlin, 361–370.

Modestino, J.W. and J. Zhang (1992) A Markov random field model–based approach to image interpretation. *IEEE Trans Pattern Anal Machine Intell*, 14:606–615.

Pan, Y. (1994) *Autoregressive Modelle für die Texturanalyse in digitalen Bildern*. Reihe C, 430. Deutsche Geodätische Kommission, München.

Woods, J.W. (1972) Two–dimensional discrete Markovian fields. *IEEE Trans Inform Theory*, IT–18:232–240.

Cartography

Semanitic Modeling for the Acquisition of Topografic
Information from Images and Maps
ed. by W. Förstner and L. Plümer

Cartographic Displacement Using the Snakes Concept

Dirk Burghardt and Siegfried Meier

Department of Geosciences, Dresden University of Technology
e-mail: burg@ipg.geo.tu-dresden.de

Abstract

Symbolized images (maps) with a very high information density have to be generalized sometimes, i.e. objects have to be displaced reciprocally. A displacement algorithm which operates by the principle of the snakes concept is described. The addition of active splines, so-called snakes, to contours (attraction) is one possiblity to recognize indefinite contours. Another one is the reciprocal displacement of lines which is exactly the inversion (repulsion) of the previous method. Due to this equivalency the snakes technique may be also applied to line displacement. The described algorithm of energy minimization provides (carto)graphically sufficient solutions. Geometric and semantic control parameters take care to optical resolution, maintenance of topological structures, minimal changes in geometry and consider the object significance.

1 Introduction

In computer graphics and in cartography sometimes symbolized images with very high information density and insufficient visual resolution or graphic constraints have to be generalized. Before starting the selection, which always means a loss of information, one should try to displace the objects in a reciprocal way as far as enough space is available. Usually, displacement problems are of hybrid nature, because geometric, topological and semantic relations between the objects to be displaced have to be considered.

Previous displacement algorithms have been working with fundamental geometric means and are still very afflicted with the cartographic way of thinking oriented to special cases and scale transitions. One exception is the spring tension model (Bobrich 1996), where an optimal principle is applied to cartographic displacement. To obtain a certain consistency between automatic recognition and elimination of graphic constraints the snakes technique can be used alternatively. Active splines (*snakes*) are added to them in an energy minimizing way (attraction) for best recognition of indefinite contours. The reciprocal line displacement is exactly the inversion of this process (repulsion); see Figure 1.

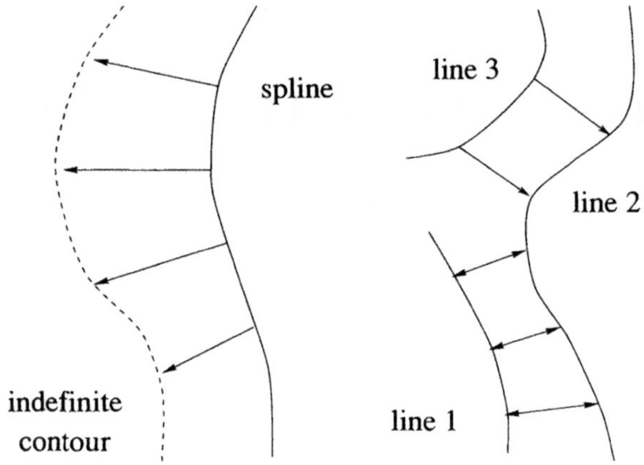

Fig. 1: Analogy between contour recognition (a spline is enclosed to an indefinite contour, left) and one-side or reciprocal line displacement (right) in the concept of energy-minimizing splines (snakes).

The equivalence of attractive and repulsive forces enables the application of the snakes technique by a variational problem also to displacement problems by minimizing the total energy. The numerical (iterative) solution algorithm, the so-called *greedy* algorithm (Williams and Shah 1990), or the stepwise solution of discretized *Eulerian* equations (Kass et al. 1987) can be used in the same way. Apart from this, the energies and potentials participating in the displacement problem should be formulated in a different way than in the recognition problem; see Table 1.

The displacement algorithm has been developed on the basis of the principle of energy minimization and tested within a first version. It provides (carto)graphically

	External Energy	Internal Energy
Recognition	Image Intensity	Direction and Curvature
Displacement	Displacement Potential	Differences in Directions Differences in Curvatures before and after Displacement

Tab. 1: Energies in the recognition and the displacement problem.

sufficient solutions. Geometric and semantic control parameters like hard-core distances which are to be given, displacement, direction and curvature potentials and weights according to object significance ensure the desired resolution. Furthermore, graphic constraints disappear whereas topological structures, fixed points and directions remain unchanged, position, direction and curvature of lines vary to a minimum extent. Finally, significant objects do not change their position, and insignificant objects are displaced.

2 Fundamentals of the Technique

2.1 The Principle of Energy Minimizing

The method of energy minimizing is applied among others to pattern recognition. It is known by the notion energy minimizing splines (snakes) (Kass et al. 1987) . In this article the method is adapted to line displacement. The question of the line course with the parameter representation $\mathbf{v} = [x(s), y(s)]^\top$ to minimize the functional

$$I(\mathbf{v}) := \int_0^1 E_{tot}\, ds = \int_0^1 (E_{ext} + E_{int})\, ds \tag{1}$$

should be the initial point. The total energy E_{tot} is integrated along the line with the arc length s, $s \in [0, 1]$. E_{tot} consists of internal and external energy. The external energy E_{ext} is used to describe the displacement cause. The short distances between the curve elements of different lines are decisive criteria. At the same time the original line shape which is described by its inclination and curvature are changed to a minimum extent. The above quantities are integrated in the internal energy E_{int}.

2.2 The Displacement Potential

The external energy E_{ext} is used to describe graphic constraints and amounts to zero if there are no displacement objects within a given minimum (hard-core) distance h from the place $\mathbf{v}_i = [x_i, y_i]^\top$. Point P_i represents an arbitrarily given point of the line L_I (see fig. 2). A displacement potential $E_{ext}(\mathbf{v}_i) > 0$ will develop in point P_i if the line or other displacement objects remain under the hard-core distance. The problem becomes more serious as the line segment extending within the hard core becomes longer and as the distance from P_i decreases. The simplest statement of the described requirements is

$$E_{ext}(\mathbf{v}_i) \sim \begin{cases} (1 - a/h) & : \quad a < h \\ 0 & : \quad a \geq h \end{cases} , \tag{2}$$

where a is the distance between the line segment performing the displacement and the investigated point P_i for which the displacement potential is to be determined:

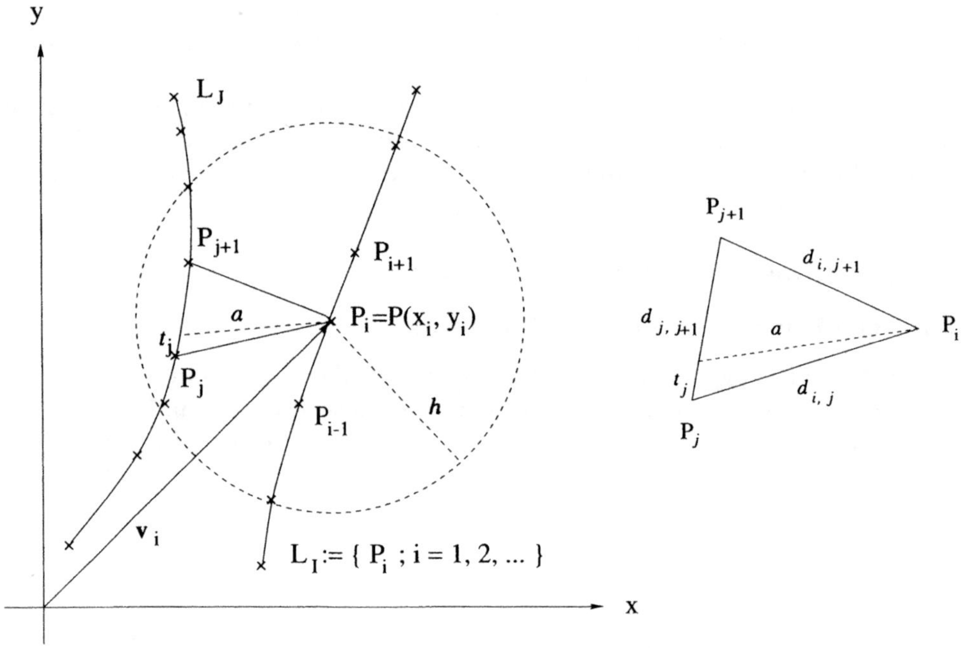

Fig. 2: Example of the determination of the displacement potential (in point P_i of the line L_I with respect to line L_J).

$$a = \sqrt{d_{j,j+1}^2 \cdot t_j^2 + (d_{i,j+1}^2 - d_{i,j}^2 - d_{j,j+1}^2) \cdot t_j + d_{i,j}^2} \tag{3}$$

with

$$\begin{aligned} d_{j,j+1} &= \sqrt{(x_{j+1} - x_j)^2 + (y_{j+1} - y_j)^2} \quad , \\ d_{i,j} &= \sqrt{(x_j - x_i)^2 + (y_j - y_i)^2} \quad , \\ d_{i,j+1} &= \sqrt{(x_{j+1} - x_i)^2 + (y_{j+1} - y_i)^2} \end{aligned}$$

(cf. Appendix A1). The calculation is performed using the sliding average. The parameter t_j results from the ratio between the step size Δ and the individual distances of the given points:

$$t_j = \frac{\Delta}{d_{j,j+1}} \quad , \quad t_j \in [0, 1) \quad . \tag{4}$$

The step size is given (fixed value) and can be chosen as an arbitrarily small value (at least one order of magnitude smaller than the average distance of the given values). Concerning $\Delta \to 0$ this corresponds to a quasi-continuous displacement.

2.3 The Shape Potential

The internal energy E_{int} is used to maintain the line shape which has been displaced due to conflicts. *Deviations* of first and second derivatives are used as quality measures (the sub-indices designate the partial derivatives by the arc length s):

$$E_{int} = (\alpha|w_s|^2 + \beta|w_{ss}|^2)/2 \tag{5}$$

with

$$\begin{aligned} w &:= (x - x^o, y - y^o) \quad , \\ w_s &:= (x_s - x^o_s, y_s - y^o_s) \quad , \\ w_{ss} &:= (x_{ss} - x^o_{ss}, y_{ss} - y^o_{ss}) \quad . \end{aligned}$$

In the iterative displacement process the differences in w_s and in w_{ss} always refer to the original line shape $(x^0, y^0, x^0_s, y^0_s, x^0_{ss}, y^0_{ss})$. In general, term (5) agrees with the internal energy used in the snakes so far. The weights α and β are constant for each line (see Section 3.2).

3 Numerical Realization

3.1 Algorithms of Energy Minimization

The Variational Calculus. Let us regard a functional

$$I[x(s), y(s)] = \int_0^1 E_{tot}\, ds = \int_0^1 E_{tot}(x, x_s, x_{ss}, y, y_s, y_{ss}, s)\, ds \tag{6}$$

of two functions $x(s), y(s)$ with fixed boundary values

$$x(0) = x_a \quad , \quad y(0) = y_a \quad , \qquad x(1) = x_e \quad , \quad y(1) = y_e \quad .$$

The functions $x(s), y(s)$ with minimized functionals $I[x(s), y(s)]$ are wanted. An inevitable prerequisite is the stationarity of I in variation of the desired functions:

$$\delta I[x + \delta x, y] = 0 \quad , \qquad \delta I[x, y + \delta y] = 0 \quad . \tag{7}$$

The Eulerian equations of the variational calculus for the two functions can be derived (Fliessbach 1992):

$$\begin{aligned} \delta I[x + \delta x, y] &= I[x + \delta x, y] - I[x, y] \\ &= \int_0^1 ds(E_x \delta x + E_{x_s} \delta x_s + E_{x_{ss}} \delta x_{ss}) \\ &= \int_0^1 ds(E_x - \frac{dE_{x_s}}{ds} + \frac{d^2 E_{x_{ss}}}{ds^2}) \delta x = 0 \quad . \end{aligned} \tag{8}$$

In this case $E_{x_{ss}} \delta x_{ss}$ develops to $(d^2 E_{x_{ss}}/ds^2)\delta x$ by partial integration done twice. The variation for the function $y(s)$ is performed in an analog way. Two differential equations of fourth order result from the arbitrariness of δx and δy, the Eulerian equations

$$E_x - \frac{dE_{x_s}}{ds} + \frac{d^2 E_{x_{ss}}}{ds^2} = 0 \quad , \qquad E_y - \frac{dE_{y_s}}{ds} + \frac{d^2 E_{y_{ss}}}{ds^2} = 0 \quad . \tag{9}$$

After the integration of the internal and external potential the following equations result:

$$\frac{\partial E_{ext}}{\partial x} - \alpha(x_{ss} - x^o_{ss}) + \beta(x_{ssss} - x^o_{ssss}) = 0 \quad , \tag{10}$$

$$\frac{\partial E_{ext}}{\partial y} - \alpha(y_{ss} - y^o_{ss}) + \beta(y_{ssss} - y^o_{ssss}) = 0 \quad . \tag{11}$$

The equations (10) and (11) are discretized by finite differences and are iteratively solved by Cholesky decomposition method (cf. Appendix A2).

The Greedy Algorithm. By the so-called greedy algorithm (Williams and Shah 1990) the energy minimization of lines can be easily performed. In this approach we attempt to reduce the energy *of each given value* by means of infinitesimal displacements. In the variational technique, however, the energy *of the whole line* is minimized in every iteration step. The transformation of conflict situations to the displacement potential (see paragraph 2.2) is the initial point of both techniques. The calculation of the internal energy is performed in a similar way. I.e. through the greedy algorithm the original line shape is registered by the explicite calculation of first and second derivatives on the given point positions :

$$E_{int} = E_{dis} + E_{curv} \tag{12}$$

with the following expressions:

$E_{dis} \sim \delta_0^2 - \delta_1^2$, while δ_0 is the original distance of the given points, δ_1 the distance of the given points after displacement,

and

$E_{curv} \sim \kappa_0^2 - \kappa_1^2$, while κ_0 is the orginal curvature, κ_1 the curvature after displacement.

E_{dis} is the energy part which results from differences in the first derivative between two iterations. These deviations can be geometrically interpreted as a change of the distance between the given points. In the analog way the contributions by variations in the second derivative are to be interpreted as curvature differences. The normalized distance between the given points results from

$$\delta = (\Delta x_i/\Delta s_i)^2 + (\Delta y_i/\Delta s_i)^2 \quad , \qquad \Delta s_i^2 = \Delta x_i^2 + \Delta y_i^2 \quad , \tag{13}$$

wherein $\Delta x_i = x_i - x_{i-1}$, $\Delta y_i = y_i - y_{i-1}$ are the coordinate differences between presently and previously given points. The curvature is calculated by

$$\kappa = (\Delta x_i/\Delta s_i - \Delta x_{i+1}/\Delta s_{i+1})^2 + (\Delta y_i/\Delta s_i - \Delta y_{i+1}/\Delta s_{i+1})^2 \quad . \tag{14}$$

Concerning discrete calculation also the approach by Williams and Shah (1990) should be compared. Furthermore, we determine the direction in which the given points are to be displaced. For this an 8-neighbourhood with a step size that should be small in comparsion to the distances of the given points can be used. If the total energy of the given point is smaller at one point of the 8-neighbourhood, this point will be accepted as a new one (see Figure 3). Since the shape potential is created only by a change of the original position of the given points (which always means an energy increase), this increase has to be compensated by a reduction of the displacement potential.

In the first step the given points of all lines should be successively treated in this way. When all given points have a minimum of energy the iteration process will be concluded.

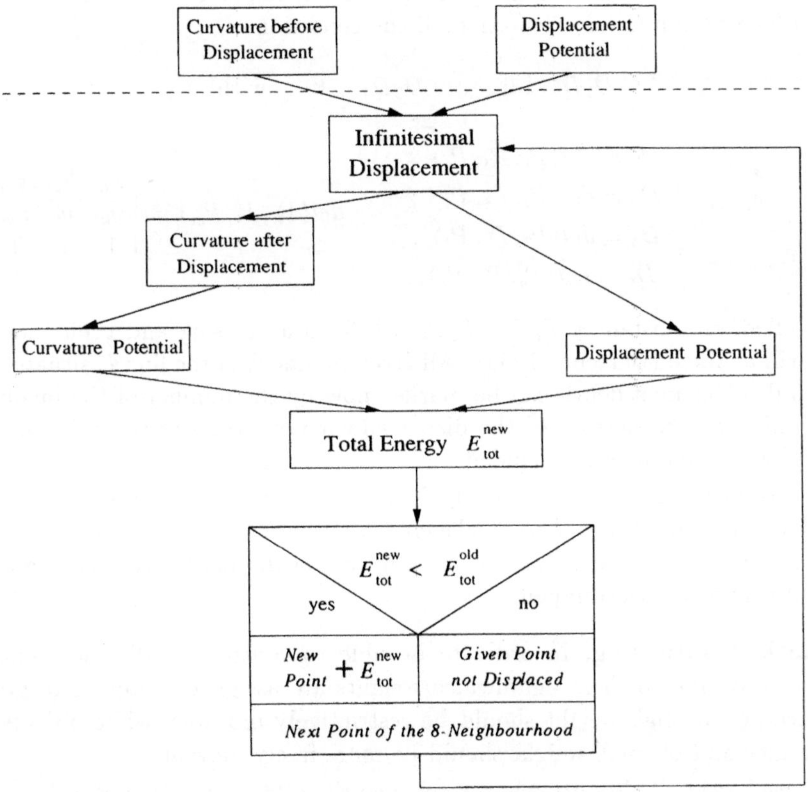

Fig. 3: Displacement scheme using the greedy algorithm.

3.2 Constraints and Controlling

Geometrical Constraints. The displacement potential is provided with a parameter to maintain the line shape around crossings. The above potential is a function of the distance between the lines, so that maximum values can be obtained in these areas. This means an orthogonalization of the (not neccessarily right-angled) crossing lines. On the one hand the mentioned parameter has to avoid the effect of orthogonalization and on the other hand a displacement of crossing lines by third objects must not be excluded. One possibility is to multiply the displacement potential by the factor

$$\chi^i(I,J) = \begin{cases} 1 & I \neq J \\ 0 & I = J \end{cases} \quad . \tag{15}$$

The superior index describes the given point for which the displacement potential has been calculated. The capital letters represent the lines.
At first the determination of the given points participated in line crossings is performed for the numerical calculation. If the conditions

$$D_1 D_2 < 0 \quad , \quad D_3 D_4 < 0 \quad \text{with} \tag{16}$$

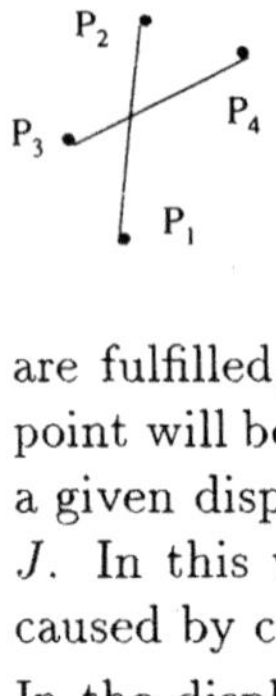

$$\begin{aligned} D_1 &= det(P_1, P_3, P_4) \ , \\ D_2 &= det(P_2, P_3, P_4) \ , \\ D_3 &= det(P_3, P_1, P_2) \ , \\ D_4 &= det(P_4, P_1, P_2) \ , \end{aligned} \qquad det(P_k, P_l, P_m) := \begin{vmatrix} x_k & x_l & x_m \\ y_k & y_l & y_m \\ 1 & 1 & 1 \end{vmatrix} ,$$

are fulfilled, the distances P_1P_2 , P_3P_4 will form a crossing and their intersection point will be fixed (Bartelme 1995). All given points P_i of the line I, situated within a given displacement depth can be marked now by the number of the involved line J. In this way an increase of the displacement potential of involved given points caused by crossing lines is excluded.
In the displacement patterns the topological relations are maintained by the procedure descrribed above. The displacement potential is defined in a way that an element can approach an adjacent one only up to the hard core, i.e. the adjacent element cannot be overjumped.

Semantic Controlling. In order to be able to control the displacement of the objects according to their significance weights are assigned. Objects of great significance and of high weight should be restrictively movable, while objects of less significance and of small weight should be more freely movable.
So far, we have only described how to change the *relative* position of objects. There are different possibilities to control the *absolute* position of lines:

i) The intensity of displacement is influenced. For that purpose the line is only partially integrated in the iterative displacement process or the step size is changed. Both procedures are controled by the significance of lines.

ii) Extension of the internal potential (5) by weighted term $\sim |w|^2$.

The first version is illustrated. According to their significance each object obtains a weight $p \in [p_{min}, p_{max}]$. In practical realization the iterative treatment of objects is used. As described earlier in Section 3.1 objects are displaced by small amounts. Before the execution of each iteration step it is decided by means of the weights, whether the object can be displaced. For a displacement the weights are interpreted as frequencies. By taking a random number z from the area of possible weights $(p_{min}, p_{max}]$ and by comparing it with the weight of the current object we can decide whether in this iteration step the object is to be displaced. The weighting of objects can be arbitrarily adapted to various applications. An object with p_{max} neither changes its position nor its shape.

Shape Controlling. The internal potential consists of two terms with the weights α and β (see Section 2.3). They control the evaluation of changed distances between given points and the deviations of line curvature during displacement. In simple cases the control for all line objects is performed with the same parameters. Larger weights provide lines with strong internal bonds.

In addition, parameters for the lines of different significance can be determined individually, the weights of the internal potential becoming semantic control parameters, too. Finally, we obtain the possibility to effect the curvature behaviour of parts of a line by using the parameters not as constants but as functions of the arc length: $\alpha = \alpha(s)$, $\beta = \beta(s)$.

3.3 Test Example

A typical conflict situation is illustrated by Figure 4(a): after transformation of the original image to a smaller scale line contacts and overlaps appear in the same symbol size. Line overlaps are displaced by energy minimization. Due to vector data (digitized line in figure 4(b)) the technique operates very fast.

As far as quality criteria are concerned the solutions from the variational technique (c) and the greedy algorithm (d) are both graphically admissible but deviate from each other, especially in the centre of the image and in the range of line crossings. The variational technique operates with fixed boundary values ("external constraint"), wherein the displacement process is stronger in the line centre than in the greedy algorithm. The mentioned effect is slightly reduced by the iterative solution of Eulerian equations. In the variational technique the typical line shape is maintained in a better way than in case of the greedy algorithm.

Sometimes the absolute position of "sharp curves" around line crossings is slightly changed. In the variational technique this applies to the upper crossing and in the greedy algorithm to the lower one. Until now these effects have not yet been eliminated and they should, with respect to the solution algorithm demonstrate, that the control parameters have to be carefully coordinated with each other (according to Section 3.2). In some cases a feed back mechanism or a trial-and-error technique should be used.

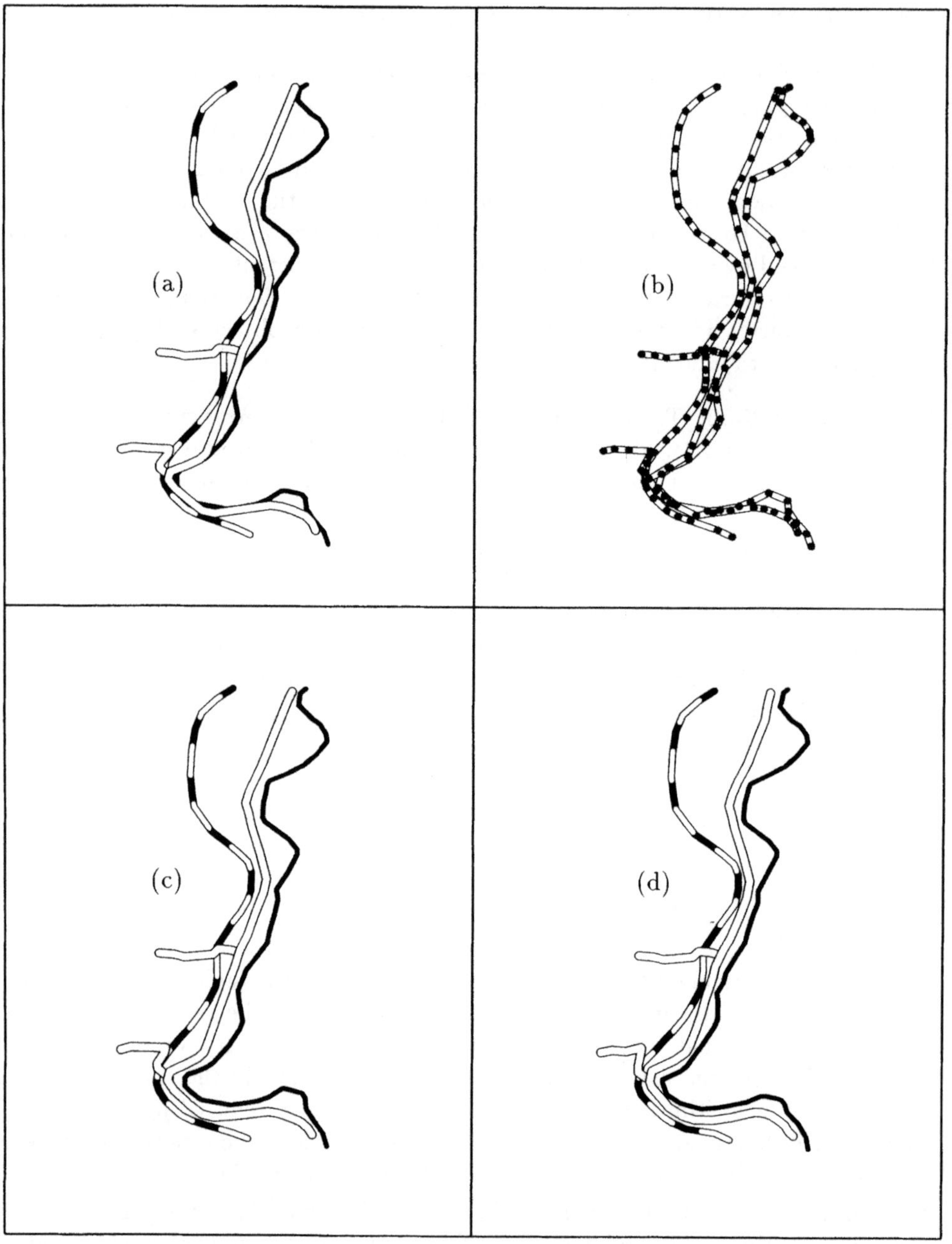

Fig. 4: Example of line displacement. Lines before displacement (a), digitized (b); after displacement by the variational technique (c) and the greedy algorithm (d).

4 Concluding Remarks

The above test example illustrates the application of the presented technique to line displacement. Point displacement by energy minimization is transformable in the same way. In the simplest case the parameters α and β of the internal potential are assumed to be zero, wherein all points are integrated in the displacement as isolated objects (point patterns). By this approach point and line displacements in mixed patterns are possible by means of the same algorithm.

So far in practical applications a linear displacement potential has been used. Other functional dependencies on the distance (quadratic progression, reciprocal distance model $1/r$) have been tested for simple examples without any visible effects. At the same time there exist different methods to calculate the curvature or the curvature potential from discrete data. They were investigated by Williams and Shah (1990) concerning their robustness.

Although the pilot version of the algorithm operates with suffcient results the problem of the adaption to practical tasks with various special cases, however, has still to be solved: definition of intersection according to data model and object structure, recognition and limitation of conflict situations, preferably by pattern recognition. The applications include large-scale and especially cadastral maps, topographic maps with coordinated scales and the transition to small scale general maps or the appertaining digital map models. Even the positioning of the lettering seems to be possible by energy minimization or by similar procedures.

Finally, we would like to underline the advantages of our computer-assisted displacement algorithm:

i) it is based on a generally valid subassigned principle, the energy minimization, therefore

ii) all special cases of graphical conflict situations in all possible data and object models, scale transitions, etc. can be solved, and

iii) it is consistent with the procedures of digital image processing.

Appendix A1:
Derivation of the distance a according to formula (3).

The starting point is the calculation of the distance between the points P_0 and P_i (Lang 1980).

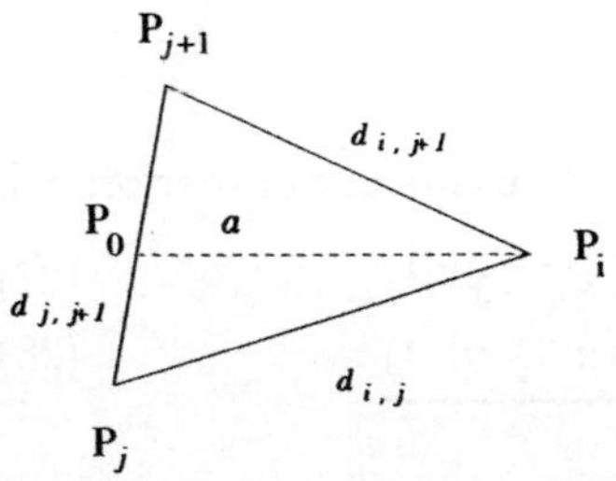

$$a^2 = (x_0 - x_i)^2 + (y_0 - y_i)^2 \qquad (17)$$

P_0 moves from P_j to P_{j+1}:

$$x_0 = x_j + \Delta x \cdot t \ , \qquad y_0 = y_j + \Delta y \cdot t \ , \qquad (18)$$

wherein

$$\Delta x = x_{j+1} - x_j \ , \qquad \Delta y = y_{j+1} - y_j \ .$$

The substitution of (18) in (17) leads to

$$\begin{aligned} a^2 &= (x_j - x_i + \Delta x \cdot t)^2 + (y_j - y_i + \Delta y \cdot t)^2 \\ &= (x_j - x_i)^2 + (\Delta x \cdot t)^2 + 2(x_j - x_i)(\Delta x \cdot t) \quad + \\ &\quad (y_j - y_i)^2 + (\Delta y \cdot t)^2 + 2(y_j - y_i)(\Delta y \cdot t) \\ &= d_{i,j}^2 + d_{j,j+1}^2 t^2 + 2t\left[(x_j - x_i)\Delta x + (y_j - y_i)\Delta y\right] \quad . \end{aligned} \tag{19}$$

With the boundary condition for $t = 1$ and $a = d_{i,j+1}$ in formula (19) we get

$$d_{i,j+1}^2 = d_{i,j}^2 + d_{j,j+1}^2 + 2\left[(x_j - x_i)\Delta x + (y_j - y_i)\Delta y\right] \quad . \tag{20}$$

Formula (3) results from the equations (19) and (20).

Appendix A2:
Discretization and solution of the Eulerian equations.

At first the equations (10) and (11) are discretized by the introduction of finite differences:

$$\begin{aligned} 0 = (E_x^i, E_y^i) + \alpha\left[(w_i - w_{i-1}) - (w_{i+1} - w_i)\right] + \\ \beta\left(\{[(w_i - w_{i-1}) - (w_{i+1} - w_i)] - [(w_{i+1} - w_i) - (w_{i+2} - w_{i+1})]\} - \right. \\ \left. \{[(w_{i-1} - w_{i-2}) - (w_i - w_{i-1})] - [(w_i - w_{i-1}) - (w_{i+1} - w_i)]\}\right) \quad . \end{aligned} \tag{21}$$

After conversion and with the substitutions

$$a := 2\alpha + 6\beta \qquad b := -\alpha - 4\beta \qquad c := \beta \tag{22}$$

the equation

$$0 = (E_x^i, E_y^i) + cw_{i-2} + bw_{i-1} + aw_i + bw_{i+1} + cw_{i+2} \tag{23}$$

results from (21), in matrix presentation

$$\mathbf{A}(\mathbf{x}^t - \mathbf{x}^0) = \mathbf{E_x}(\mathbf{x}, \mathbf{y}) \quad , \tag{24}$$

$$\mathbf{A}(\mathbf{y}^t - \mathbf{y}^0) = \mathbf{E_y}(\mathbf{x}, \mathbf{y}) \quad , \tag{25}$$

with the pentadiagonal band matrix

$$\mathbf{A} = \begin{pmatrix} a & b & c & 0 & 0 & \cdots \\ b & a & b & c & 0 & \\ c & b & a & b & c & \\ 0 & c & b & a & b & \\ 0 & 0 & e & d & a & \\ \vdots & & & & & \end{pmatrix} \quad . \tag{26}$$

In the next step the transition to an iterative treatment (parameter t) is performed:

$$(\mathbf{A} + \gamma\mathbf{I})(\mathbf{x}^t - \mathbf{x}^0) = \mathbf{x}^{t-1} - \mathbf{x}^0 - \mathbf{E_x}(\mathbf{x}^{t-1}, \mathbf{y}^{t-1}) \tag{27}$$

$$\underbrace{(\mathbf{A} + \gamma\mathbf{I})}\underbrace{(\mathbf{y}^t - \mathbf{y}^0)} = \underbrace{\mathbf{y}^{t-1} - \mathbf{y}^0 - \mathbf{E_y}(\mathbf{x}^{t-1}, \mathbf{y}^{t-1})} \quad . \tag{28}$$

The vectors $\mathbf{x}^t$ and $\mathbf{y}^t$ represent the x- and y-coordinates of the line in the present iteration process. $\mathbf{x}^0$ and $\mathbf{y}^0$ are the x- and y-coordinates of the original line. The solution of the decoupled matrix equations (27) and (28) is realized by the Cholesky decomposition method. It is a symmetric version of the LR decomposition for positive definite matrices:

$$\begin{aligned} \mathbf{Bn} &= \mathbf{m} \\ \mathbf{R}^{\mathbf{T}}\mathbf{Rn} &= \mathbf{m} \\ \mathbf{R}^{\mathbf{T}}\mathbf{u} &= \mathbf{m} \quad \Rightarrow \quad \mathbf{Rn} = \mathbf{u} \quad \Rightarrow \quad \mathbf{n} \end{aligned} \tag{29}$$

References

Bartelme N. (1995) *Geoinformatik - Modelle Strukturen Funktionen*, Springer-Verrlag, Berlin Heidelberg.

Bobrich J. (1996) *Ein neuer Ansatz zur kartographischen Verdrängung auf der Grundlage eines mechanischen Federkraftmodells*, Deutsche Geodätische Kommission, Reihe C, H. 455.

Fliessbach T. (1992) *Mechanik*, Wissenschaftsverlag, Mannheim, Leipzig, Wien, Zürich.

Kass M., A. Witkin, D. Terzopoulos (1987) *Snakes: Active contour models*, Proceedings of the First International Conference on Computer Vision, IEEE Comput. Soc. Press, pp. 259–268.

Lang W. (1980) *Gleitende Flächen mit strenger Berücksichtigung von Bruchkanten*, Master's thesis, Univ. Bonn.

Williams D. J., M. Shah (1990) *A fast algorithm for active contours and curvature estimation*, Proceedings of the Third International Conference on Computer Vision, IEEE Comput. Soc. Press, pp. 592–595.

Acknowledgement

The authors, who have not yet been involved in image processing so far, owe *Wolfgang Förstner* (Bonn) a debt of gratitude, because has drawn our attention to the excellent possiblities offered by the snakes technique. We also thank *Frank Bethge* (Dresden) who has assisted in our work in a critical and helpful manner.

Semanitic Modeling for the Acquisition of Topografic
Information from Images and Maps
ed. by W. Förstner and L. Plümer

Multi-level Comparative Analysis of spatial Operators in GIS and Remote sensing as a Foundation for an Integrated GIS

Stefan Jung, Jochen Albrecht und Manfred Ehlers

ISPA, University of Vechta, Germany
P.O.Box 1553, D-49364 Vechta
Ph.: +49-4441-15 337, Fax: +49-4441-15 445
e-mail: [sjung, mehlers] @ispa.uni-vechta.de, j.albrecht@auckland.ac.nz

Abstract

In spite of many attempts, the integration of GIS and image processing leaves much to be desired. One explanation is the rather simplistic approach of adding functionality without acknowledging the different semantics implicit in any operation. A hierarchical decomposition of common remote sensing tasks provides the key to a more sophisticated integration at *any* stage of the image analysis. Depending on the scale at which reasoning takes place, different forms of context come to look in: regular attributes, meta information, or real knowledge bases. The formalization of the context for each operation precludes any structural or semantic misunderstanding, True interoperability, however, can only be achieved if low-, mid- and high-level operators not only match each other in the GIS and image processing domain but also sustain a free flow between data models. Quality measures keep track of the error introduced, and serve as decision support for further processing steps.

1 Introduction

While Integrated GIS (IGIS) has been the subject of research initiatives (Star et al. 1989, Ehlers et al. 1989), comments (Dobson 1993, Wilkinson 1996) and research (Jackson and Mason 1986, Woodsford 1994) for some years, users who are actually attempting to practice integration have had to use inappropriate software, converting formats and transferring datasets between systems (Trotter 1991). A scientist who might wish to update existing cadastral (land registry) information in a vector format with satellite data to combine both in analysis, poses the minimal set of requirements that a system must fulfill to deserve the name Integrated GIS. The general intention of all these attempts is to overcome obstacles using image processing algorithms and remote sensing techniques in combination with GIS functionality, a concept which will be refined in section 5.
Image processing as well as spatial data analysis systems and GIS all consist of very complex bundles of algorithms making it very hard for non experts to get an overview of

how to use them as a tool for the accomplishment of certain well defined tasks. They both serve dedicated communities and their set of functionalities is rather diverse. All image processing systems are raster-oriented and hence more closely related to raster-based GIS than to their vector-based cousins. Even here, however, the application domains differ: GIS tend not to offer multispectral analysis whereas very few image processing systems are able to analyze elevation data. Vector capabilities are usually veneer. Integration, so far, has in most cases been restricted to "live links" between different systems that allow to read native formats rather than to import data. A truly Integrated GIS, however, would include the functionalities for both domains within a single package (see figure 1), although from a software engineering point of view, this monolithic appearance is internally replaced by a shell that supports a variable number of plug-ins (see sections 4 and 6).

Current research in interoperability (OGC 1996, ISO 1996) suggests that there are a number of low- and mid-level spatial operators that can be employed by both image processing systems and GIS. An Integrated GIS could then use the same algorithm in either application and therefore reduce the bulk of redundant functions. This requires a comparative examination of each operation, a task which can only be accomplished on a formalized basis. After a formalization of GIS operations, the authors are presently engaged in a comparable effort for image processing operations.

An Integrated GIS should support the convergence from both sides, i.e., it should facilitate the use of ancillary GIS data within image interpretation as well as the use of remotely-sensed data within GIS analysis. In the following, the authors demonstrate a showcase of how the integration of distributed remote sensing and image interpretation algorithms can be accomplished within the Virtual GIS (VGIS), a graphical user interface for environmental scientists who are non-expert users of GIS (Albrecht *et al.* 1996). Section 3 discusses the possible benefits of an integration of those two domains in environmental applications. Section 5 introduces a hierarchical organization of selected image analysis operations similar to the likewise systematization of GIS operations presented in Albrecht (1996). This classification is limited to the analytic capabilities needed in geoinformation (as opposed to architecture or industrial applications), specifically multispectral analysis and classification as well as georeferencing. Section 0 examines software engineering requirements, particularly the concept of a method warehouse, operation registry, and the use of metadata as a mean to introduce some intelligence that supports non-expert users of an Integrated GIS. Special emphasis is given to the continuos accompaniment of quality measures and error propagation. The concluding section 7 relates this work to the endeavors of the Open GIS Consortium and discusses future research needs.

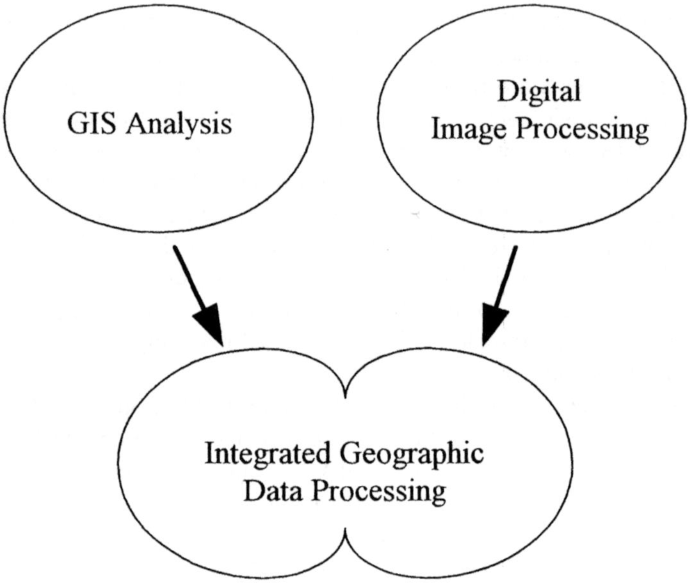

Figure 1: Development of an Integrated GIS.

2 Current Developments in Remote Sensing

Three methodological and one technical advancement have a profound influence on the nature of remote sensing as it is known so far (Ehlers 1995a, 1995b). One is the emergence of high-resolution (1-3 m) data leading towards completely new application areas, e.g., large-scale mapping. Since most GIS data capture cadastral and other rather detailed information, the advent of systems such as Early Bird, Quick Bird, or Orb View-1 (Fritz 1996, Nale and Thurgood 1996) will, for the first time, provide operational co-use of vector and raster data of the same native resolution. At the same time, the traditional classification methods are expanded by sophisticated texture measures, neural network techniques and the development of knowledge bases.

Spatial pattern analysis has so far been a domain of the spatial analysis modules of larger GIS. Recent work at the Joint Research Centre - Institute for Remote Sensing Applications (ISPRA, Italy) (Wilkinson *et al.* 1995) describes multifractal models as a suitable tool to calculate texture measures to be utilized along with spectral data classification. Rather than developing new methods for image processing software, integration with GIS would make use of the respective statistical modules inherited from GIS (Pyka and Steinnocher 1994).

Particularly in environmental applications (Goodenough et al. 1987, Lees and Ritman 1991), neural networks have proven valuable in the classification of cluster datasets. While the sole use of neural network classifiers is usually not more successful than traditional methods, its combination with more determined (vector) data can prove to be especially effective after the GIS has been used to stratify an image. A common difficulty lies in the selection of the appropriate initial configuration (Gahegan et al. 1996). Again, a GIS provides the necessary "knowledge" about the feature space to make an educated guess for the setup of a neural network.

Compared with the comparatively raw data processed in remote sensing, any GIS can be regarded as a knowledge base, i.e., the attributes that usually accompany the geometric data deliver sufficient ancillary data to serve as a knowledge base. This is, however, not the notion employed here. Knowledge-based systems, in their proper sense, provide mechanisms for the automatic derivation of information from data. Early work concentrated on simple heuristics such as the modeling of pest infestation (Elmes and Cay 1992) or regularly shaped (man-made) features in aerial imagery (Fischler et al. 1981, Matsuyama 1987). A more advanced integration employs a full-fledged expert system that steers the execution of each individual GIS or image processing function (Miller 1994). Albeit operational (the system models vegetation change for all of Scotland), this system is an intricately woven structure that is nowhere close to commercialization, but it demonstrates a feasible direction for Integrated GIS.

3 Benefits of the Integration of GIS, Remote Sensing and Image Analysis

As annunciated above, the advantages go both ways, i.e., digital processing of remotely sensed images can benefit from the integrated use of analytic GIS operations and many GIS applications depend on remotely sensed data and hence would benefit from image lineage information associated with objects. It has been suggested that the synergism between remote sensing and GIS is the major advantage of an integration supplying more functionality than the mere sum of the functions of an image analysis software and a GIS (Ehlers 1989, Star *et al.* 1991, Trotter 1991). The following examples may illustrate this point:

- Classification can be improved through the inclusion of ancillary data as another image band (Hutchinson 1982). True integration of GIS and image processing systems optimizes the extraction of information from imagery by utilizing all of the related data and functionality that the GIS can provide.
- Remote sensing analysis is increasingly using GIS data. Examples of this include overlay of vector data for presentation and location (De Cola 1989), and the use of digital elevation models (DEMs) for terrain correction of imagery and perspective visualization. High resolution topographic data make a very significant contribution to the interpretation of radar image data (Kwok *et al.* 1987). DEMs are also used for geometric correction of optical remotely sensed data (Bertolo *et al.* 1993). Conversely, the generation of DEMs directly from remotely-sensed data (e.g., interferometric SAR or altimeters) appears very promising (Trotter and Dymond 1993).
- Modern image processing methods no longer need to work just in pixel mode but make use of areal groupings. This invites a combined analysis not only with raster-based GIS but also with their vector-based cousins.

- GIS-derived data is increasingly used to post-process the classified image (Gahegan 1995, 1996, Gahegan et al. 1996). The intermediate results and statistics from image processing can be stored as attributes of polygonal data. This supports qualitative decision-making and can facilitate change detection without performing the classification of multi-temporal datasets (Derenyi *et al.* 1992). In an integrated GIS, it is a simple operation to generate statistics within polygons and then return those statistics directly to the GIS database attributes of polygons (see Figure 2).

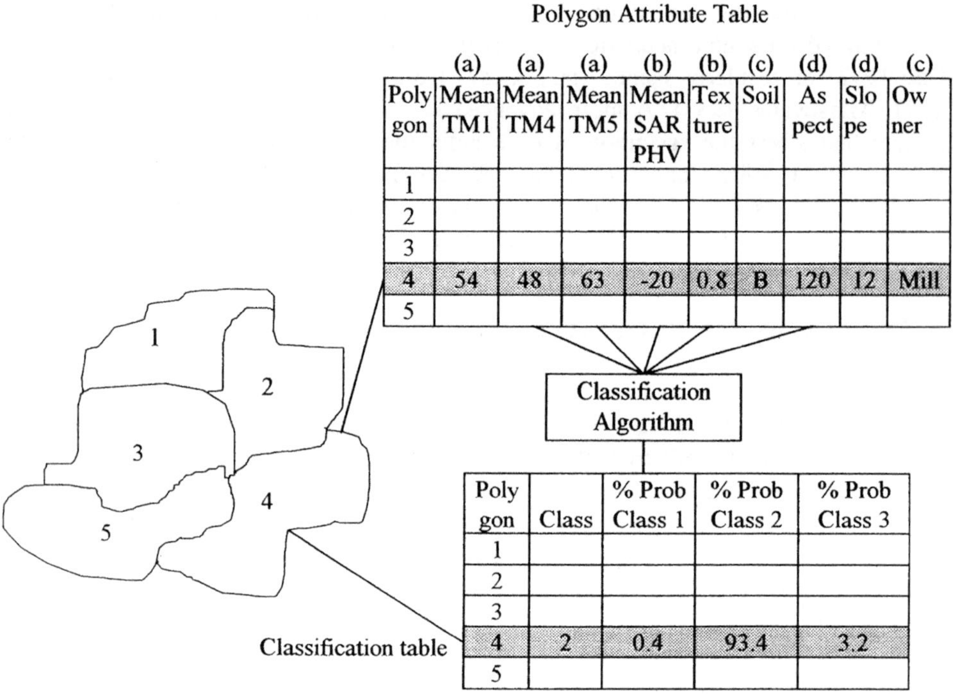

Figure 2: Schematic representation of polygon classification functionality. Attributes stored in a database table are used in the classification, rather than values of pixels in a series of image layers. Attributes could be derived from images, e.g., (a) mean value of TM bands with each polygon, (b) mean SAR backscatter signal or image texture over the area defined by the polygon, (c) typical attributes attached to the vector dataset, or (d) values obtained from a DEM within each polygon. The output of the classification is in the form of a database table which can be referenced to the vector dataset and the original attribute table, through the unique polygon identifier (Figure and Caption from HINTON 1996, p. 883).

Three types of integration can be distinguished:

(1) Projection of information from remotely sensed data to objects and the other way around,

- e.g. query of the statistics of a remote sensing data set within a certain polygon of a vector layer, without the need to transform between data formats.
- transfer of attributes from objects to spatially related raster cells or image sections (input of ancillary information).

(2) Combined analysis between all different kinds of geodata,

- e.g. the use of vector objects to limit image processing algorithms geometrically (maybe even with explicit priorities depending upon certain attributes of the vector layer)
- creation of new data sets as a result of an overlay between raster and vector information, where certain constraints can be placed on the raster layer as well as on vector data.

(3) A metadata (knowledge) controlled conversion of raster data to vector format and vice versa. That's a point, where the discrepancy between field and object view (Couclelis 1992) comes into perspective. Generally, information of different qualities should each be kept in their most adequate digital representation, at least according to the desired analytical results (even if it is not a spatial one). That does only make sense as long as the transfer of information between differently represented sources of input/data storage is warranted. So far, different representations are often separated by the incompatibility of systems as well. An integrated GIS allows

- the lineage of a data set to be stored and retrieved (see section 6).
- the algorithms of the conversion to be selected by the user.

Yet the technical approach as an improvement of functionality is only the low-level foundation for a real integrated system. At least as important is the interface between system and user (Frank 1993). The notion of *interface* here includes:

- services, i.e., functionality included in algorithms and operators, as well as
- the user friendliness in structure and supply of the actual functionality implemented under an intuitive GUI.

4 Technical Realization of Integration (VGIS)

The technical approach is to combine functions from different domains (e.g., GIS and image processing) under one GUI, allowing a seamless interaction between remote sensing and GIS operations. As it has been the case under the solely GIS-centered VGIS prototype (Albrecht et al. 1996), the emphasis on the extended VGIS model rests on analytic operations. Services can be defined by the user with a graphical flow chart editor, combining low-level operations. Each flow chart then represents the semantic model as seen by the user. Since implementation is to be realized within a single environment, both image processing and GIS operations are at the user's discretion. This implies that for the first time, a remote sensing analyst may utilize GIS operations during the definition of image processing services. Each VGIS flow chart is a singular model

(with a given methodology) that may become part of a larger shared library, e.g., of different classification algorithms. This results in freely exchangeable and combinable operations that address all three of the above mentioned types of integration.
Following the well-tried strategy of the VGIS project, operators should not be rewritten, but employ whatever is offered by existing systems. The technical realization relies on the distributed application programming interface (API) of the plugged-in software.
Modern object-oriented software engineering principles (Rumbaugh et al. 1991) tend to prefer applet-based solutions with special sets of needed functions and algorithms for a certain task over a system that meets all desires of remote sensing and GIS functionality (fatware). Systems become pluggable (see also section 0) and allow the user to 'rent a function' for a certain amount of time even if it is used only once in an application's life. The realization of this concept, however, requires further standardization in information technology (ISO 96).
Albrecht and Kemppainen (1996a, 1996b) describe the advantages of a hierarchical structuring of geospatial operations in the context of current standardization efforts, assembling groups of operators to higher level (domain specific) services.

5 The Hierarchical Structure of Selected Image Analysis Operations

The integration of the whole spectrum of image interpretation operators makes a GIS not only more flexible with respect to data input and analysis, but also more complex. With growing diversity it becomes increasingly important to structure the embedded functions according to the knowledge and preference of the user. The operations have to be composed and named in a task-oriented way. Albrecht (1996) emphasizes the user friendliness at a high level in his formal definition of 20 universal analytical GIS operations. Based upon these operations, VGIS has been designed as a GUI for non GIS-experts.
With the integration of image analysis of remotely sensed data, the problem of universal accessibility for the inexperienced user becomes convoluted. Extraction of information from remote sensing data generally requires too much knowledge to generalize it by just combining existing low-level operations and algorithms (Gahegan and Flack 1993). Depending on the data source and resolution in combination with the kind of feature to extract, individual models of different complexity have to be designed.
Albrecht (1996) demonstrates that GIS operations are easier to use for non-GIS experts and can yet be differentiated to fulfil the needs of many scientific communities, if the operations are structured hierarchically. The authors claim the same to be true if applied to image processing, i.e., the full variety of functions today's image processing tools supply is much easier to access in a hierarchicaly organized user interface[1]. Hierarchical

[1] For example before actually classifying a satellite image, various procedures may have to be applied such as radiometric normalization, geometric correction, noise filtering and more (Wilkinson 1993). For most people, who just want to get a certain result out of temporal different images, it is not of major interest how these processes work and exactly which algorithms are applied. In these cases it makes sense to agglomerate all the single low or mid-

reasoning implies also, however, that users may choose the level of abstraction; do they want to extract an edge, a polyline, or a street ? Hence it is the pronounced interest to include auxillary data which usually comes from GIS. It very much depends on both, the data source and the application to decide which method / function / operator to use (Gahegan 1996). For the problem of growing diversity through the large number of necessary individual configurations no quick solution is at hand, but a metadata registry as described in section 0 is an instrument that can supervise the compound procedures.
The argument used here works with prototypical examples frequently employed in the domain of environmental monitoring such as landuse classification with multispectral satellite data, and georeferencing from satellite images (see Table 1).

domain	GIS	Remote Sensing / Image Processing
high-level	(application) Services (OGC 96)	*information retrieval* landuse classification, object extraction (streets, buildings) *georeferencing* automatic geo-referencing with error control mechanisms,
mid-level	20 universal GIS operations (Albrecht 96)	*information retrieval* (multi-)spectral analysis, segmentation, regionalization, classification, histogram matching, histogram equalisation, contrast stretch, training area selection *georeferencing* control point detection, transformation and resampling methods
low-level	elementary spatial operators (ISO 96)	*information retrieval* filter techniques: compass, convoluted, gauss, hipass, laplace, line, lopass, Sobel, Kirsch, edge detection algorithms: NDVI, FFT, calculation of covariance, variance matrices. *Georeferencing* General data management functions: input/output

Table 1: Hierarchical structure of selected RS and GIS operations

level procedures needed to just one application specific operator to be applied every time the user analyses a new image with respect to his application specific designation (e.g. landuse classification).

6 The Multifunctional Geodata Registry (MGR) - A Register for Data, Methods and Applications

The described complexity of all needed operations as well as the handling of many different data formats and applications requires advanced management capabilities in integration. In many cases information about accuracy and history of available data to use in analysis and cartographic presentation is a major issue. The construction of a multifunctional geoinformation registry (MGR) is an appropriate method to realize integration on a technical level.

This section describes the different kinds of information stored and what such a prototype of management system could look like. The registry has a central steering function for all operations demanded by the user (see figure 3). To construct such a registry effectively and free of redundancy, the database schema needs to be formalized and supplied as the first step in an implementation.

Management functionality provided by the registry relates to the following subjects:

- registration of low-level operators, mid-level software-systems and high-level services
- metainformation,
- intelligent data handling, including methods as well as quality measurement and error propagation.

The MGR should be structurally object-oriented to enable conformity not only with modern database schemas, but also to be compatible with the OGC (1996) model in progress. Object-orientation also enables a different representation of geographically relevant data, no matter if objects or phenomena, related to different user perspectives or application goals (Couclelis 1992, OGC 1996b).

6.1 Registration of Accessible Services

The technical core for the integration of different systems acts like a coordinating central office, where every object has to have an entry. As for now (1996), this architecture needs procedures to identify software packages, operations as part of a proprietary system and their data formats as well as the system's export functions. Flexibility is a major feature of the registration system. At the current state, the registry only simulates those desirable future conditions. Nevertheless a systematic organization of relevant software systems and a digital catalogue of the functions implemented are the first steps towards a user oriented level of an integrated GIS. Last but not least, even an indexed collection of relevant algorithms in the systems on the market would be a progress. The ultimate goal is the development of an integration level that goes completely conform with the demands of the OGIS specification (OGC 1996b).

All additional features have to be declared in a separate routine, which checks the compatibility to existing objects as well as algorithms to choose from, for instance how to convert the new data-type to other formats. Minimal requirements about the depth of information can be defined in registration schemata to allow new entries only with the appropriate system knowledge.

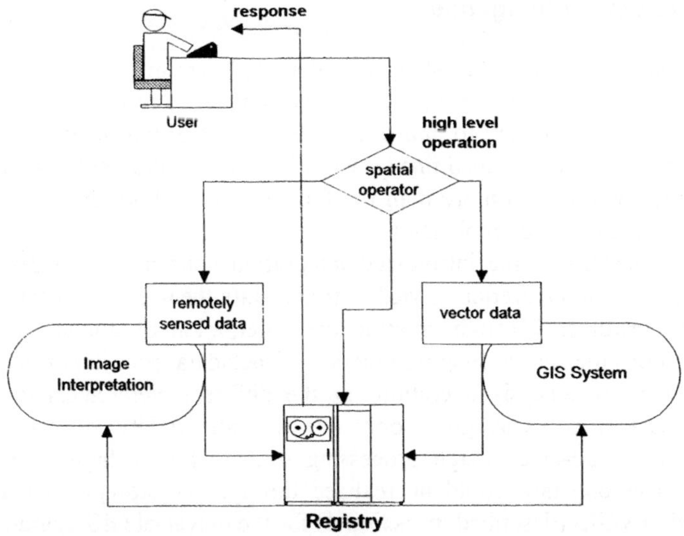

Figure 3: An example for the principle connection between the user and the registry managed system. The user contacts a spatial operator of a free to choose functional level (mostly high-level), the registry takes over organization and the triggering of functions out of different underlying systems and finally gives the O.K. for submission of results.

6.2 Metadata

Information about the accessible data is realized in the classical metadata concept (Ganter 1993) as part of the MGR. The original resolution and the thematic contents of a data layer for instance are decisive for the granularity of analytic results. The registry decides upon the metainformation which datasets can be combined, which new metainformation will be attached and of what data-format the output will be.
Unique inheritance of metainformation has to be guaranteed. The history administration of the MGR stores every step a dataset goes through, from the beginning (digitizing or scanning) over all operations already performed on the data to the end (result or deletion). The higher the hierarchical level of abstraction or aggregation of spatial operators gets, the more important becomes a consistent metainformation management (see section 5).
Combined queries enable the user to get a good overview over the data contents according to thematic detail and general attributes, e.g. actuality, as well as the more technical information about the system in which the data were acquired and about the services available. The object oriented approach principally enables the encapsulation of methods together with data. The data needed for a service demanded by the user are to be checked for their eligibility, before they can be released for processing.

6.3 The System's Intelligence

One of the major goals in establishing an integrated system is an enhancement in the effective use of remotely sensed data. Ancillary knowledge in form of existing geodata on the other hand is necessary to improve the extraction of information out of remotely sensed data. Both steps combined finally will offer users of different domains (especially in the environmental sector) more than the full capacity of both instruments to achieve better results in their field of application.

One of the challenges of the introduced integration concept is to offer an approach towards realization of integration as well as to facilitate the use of applications that are in need of both features. Through creation and integration of domain and application specific operations the central steering function of metadata gains importance.

The GIS and image analysis operations at the different hierarchical levels described above (see section 5), will give room for a wide variety of combinations. An agglomeration of mid-level image processing operators to a high-level classification operator for a certain task could be realized through the storage of a flow chart as demonstrated in VGIS (described in section 4) for the universal GIS operations.

The controlling function of the registry regarding complex compound operators has to check all mid-level operations contained within a flow chart with respect to the supplied data, the quality of the data and the requested methods.

Such a controlling mechanism can work on two different levels:

- the nominal checks make sure that the technical details fit together, that operators are implemented, that the operations can be performed on the input data, that the compatibility between different data formats can be guaranteed.
- The internal consistency registration must control the formal correctness of the performed operations within a flow chart. Formal errors such as always empty sets as a result or contrary functions within an flow chart should at this point be found.
- Thematic contents of data sets demand in a more complex way metadata induced decision mechanisms. Application specific standard combinations of mid- and low-level operators, such as certain classifiers and filters will be stored in flow charts. Together with constraints on thematic and/or geometrical attributes of the input data these flow charts have to be analyzed and evaluated. Before a high-level flow chart can be launched by the end-user, a metadata triggered evaluation result gives an estimation of the expected output quality.

  For these features the registry has to offer an environment flexible enough to be suitable for a wide spectrum of domains.

That finally leads to a kind of decision support about the desirable and acceptable accuracy of results in relation to the quality of the input data. A clue can also be given towards methods and suggestions of data acquisition necessary to complete a chosen task, e.g., a detailed classification of fen exploitation areas (operation).

Ehlers and Shi (1996) have developed a framework for handling uncertainties for the integration of GIS and remote sensing using an analytical approach. As a first step, an analytical model for combining positional and thematic errors in the integration of a classified remote sensing image data into a GIS layer was developed. Although this model is currently restricted to point mode digitizing as the sole positional GIS error

source and the maximum likelihood probability vectors as the sole classification error descriptors, it already provides quantitative results that prove the validity of considering both, positional and thematic error, for the uncertainty modeling process.
According to the results presented in the paper, we can state that it is feasible to implement procedures for providing uncertainty related information to analysts and decision makers. Shi (1994) has already developed a visualization component that can relate this information as a graphical output to the GIS user. We believe that uncertainty modules such as these will be a necessary and critical component for integrated GIS and image processing systems in the future.
As a next step, the developed model will be extended to include more than two input layers. Shi and Ehlers (1996) have already used it for a change analysis based on multiple remote sensing data. With the ability to handle more than two input errors, it will be possible to extend this model to the combination of multisource datasets in an integrated GIS environment (Shi 1994).
There are, however, a number of problems that need to be addressed in further studies. A critical step will be the inclusion of correlated point errors which occur in streamline digitization or during automated data capturing. In addition, the modeling of temporal uncertainties will be a critical problem in change detection. Another important issue is the problem of uncertainty in topological relation. Last but not least, we have to develop practical implementation strategies so that a GIS can still do what it is designed for: perform spatial analyses without being overburdened by cumbersome error assessment strategies.

7 Conclusion

When the term Integrated GIS (IGIS) was coined, the authors (Star et al. 1989, Ehlers et al. 1989) couldn't anticipate the developments leading towards the latest buzz interoperability. The user (or vendor) driven efforts in the standardization arena have such a profound impact that the implementation of IGIS will look a lot different from what was envisioned before. An interoperable IGIS, however, has to meet the user at any semantic level and should offer its range of functionalities at every stage of data (and later information) processing. The same way that humans reason at various semantic levels concurrently (Rasmussen 1986), the user needs access to low and high level analysis in any phase of IGIS usage. The reader is invited to join the authors in their efforts to

- develop a concordance of GIS and image processing terms,
- formalize GIS and image processing operations at all levels (a lot has been done here on the GIS side already (Albrecht 1996, OGC 1996, ISO 1996)),
- provide formalizable work flows from real world applications.

The authors are simultaneously developing the specification of such a network of GIS and image processing operators and implement it as an extension of the VGIS.

Literature

Albrecht, J. (1996) *Universal Analytical GIS Operations: a task-oriented systematization of data structure-independent GIS functionality leading towards a geographic modeling language*. Dissertation, ISPA Mitteilungen 23. University of Vechta: Vechta, Germany.

Albrecht, J. and H. Kemppainen, (1996a) *A Framework for Defining the New ISO Standard for Spatial Operators*. Proceedings of the GIS Research UK 1996 Conference (GISRUK'96), 10-12 April 1996, University of Kent at Canterbury, pp. 78-84.

Albrecht, J. and H. Kemppainen, (1996b) *Scalability of Spatial Operators*. Proceedings GIS/LIS'96, 18-21 November 1996 Denver CO, pp. 162-172.

Albrecht, J., H. Brösamle and M. Ehlers, (1996) *A Graphical Front-End for User-Oriented Analytical GIS Operations*. Proceedings XVIII ISPRS Congress, XXXI, B2, Commission II, Vienna, Austria, pp. 78-88.

Bertolo, F., S. Folving, S., J. Megier, J., M. Paracchini, and A. Toccolini, (1993) *Application of remotely sensed data and GIS for environmental management in an Alpine valley—a case study from Val Malenco, Italy*. In Proceedings of the International Geoscience and Remote Sensing Symposium, (IEEE: Tokyo), pp. 2065-2067.

Couclelis, H. (1992) *Beyond the Raster-Vector Debate in GIS*. In FRANK, A., CAMPARI, I. and U. FORMENTINI(Eds.) Theories and Methods of Spatio-Temporal Reasoning in Geographic Space. Berlin, New York, Springer. 639, pp. 65-77.

De Cola, L. (1989) *Fractal Analysis of a Landsat Scene*. Photogrammetric Engineering and Remote Sensing, 55(5), pp. 601-610.

Derenyi, E., R. Pollock, and D. Fraser (1992) *Training statistics generation and image classification in a GIS environment*. Proceedings of the 15th Canadian Symposium on Remote Sensing, (CASI: Ottawa), pp. 474-477.

Dobson, J. (1993) *Commentary: A conceptual framework for integrating remote sensing, GIS, and geography*. Photogrammetric Engineering and Remote Sensing, 59, pp. 1491-1496.

Ehlers, M. and W. Shi, (1996). *Error Modelling for Integrated GIS*, Cartographica (in press)

Ehlers, M. (1987) *Integrative Verarbeitung von digitalen Bilddaten der Satellitenphotogrammetrie und -fernerkundung im Rahmen von Geographischen Informationssystemen*. Wissenschaftliche Arbeiten der Fachrichtung Vermessungswesen der Universität Hannover: 137.

Ehlers, M. (1993) *Integration of GIS, Remote Sensing, Photogrammetry and Cartography: The Geoinformatics Approach.* Geo-Informations-Systeme 6(5), pp. 18-23.

Ehlers, M. (1995a) *The Promise of Remote Sensing for Land Cover Monitoring and Modeling.* Proceedings Joint European Conference and Exhibition on Geographical Information, The Hague, The Netherlands, Vol. **2**, pp. 426-432.

Ehlers, M. (1995b) *Availability of Core and Baseline Data for Global Mapping: GIS to the Rescue?* In Proceedings Cambridge Conference for National Mapping Organisations, St. John's College, University of Cambridge, 13 p.

Ehlers, M. and U. Rhein (1996) *The Role of Remote Sensing and GIS for an Operational State-Wide Environmental Monitoring.* Proceedings XVIII ISPRS Congress, XXXI, B4, Commission IV, Vienna, Austria, pp. 684-689.

Ehlers, M., G. Edwards and Y. Bédard (1989) *Integration of Remote Sensing with Geographic Information Systems: A Necessary Evolution.* Photogrammetric Engineering and Remote Sensing 55 (11), pp. 1619-1627.

Elmes, G. and G. Cay (1992) *Data Quality Issues in User Interface design for a Knowledge-Based Decision Support system.* Proceedings of the 5th International Symposium on Spatial Data Handling, Charleston, SC: International Geographic Union, Vol. 1, pp. 303-312.

Fischler, M., J. Tenenbaum and H. Wolf (1981) *Detection of Roads and Linear Structures in Low-Resolution Aerial Imagery Using a Multisource Knowledge Integration Technique.* Computer Graphics and Image Processing, 15, pp. 201-223.

Frank, A., (1993) *The use of GIS: the user interface is the system.* In Medyckyj-Scott, D. and H. Hearnshaw (Eds.) Human Factors in Geographic Information Systems, London: Belhaven, pp. 3-14.

Fritz, L. (1996) *The Era of Commercial Earth Observation Satellites.* Photogrammetric Engineering and Remote Sensing **62**(1), pp. 39-45.

Gahegan, M. (1993) *The Use of Context Within a GIS.* Proceedings of the 21st Annual AURISA Conference, Adelaide, vol. 1, pp. 161-167.

Gahegan, M. and J. Flack (1993) *Query-Centred Interpretation of Remotely-Sensed Images within a GIS.* Proceedings of the Fourth European Conference and Exhibition on Geographical Information Systems, EGIS '93, Vol. 2, pp. 942-951.

Gahegan, M. (1995) *Proximity Operators for Qualitative Spatial Reasoning.* In Frank, A. and W. Kuhn (Eds.) Spatial Information Theory, Lecture Notes in Computer Science 988, Berlin: Springer, pp. 31-44.

Gahegan, M. (1996) *Specifying the Transformation Within and Between Geographic Data Models.* Transactions in GIS 1(2).

Gahegan, M., G. German and G. West (1996) *Automatic Neural Network Configuration for the Classification of Complex Geographic Datasets.* Proceedings First International Conference on Geocomputation, Vol. 1, University of Leeds, pp. 343-358.

Ganter, J. (1993) *Metadata Management in an Environmental GIS for Multidisciplinary Users.* Proceedings GIS/LIS '93, 1, pp. 233.-245.

Gernazian, A. and S. Sperry (1989) *The Integration of Remote Sensing and GIS.* Advanced Imaging.

Goodenough, D., M. Goldberg, G. Plunkett and J. Zelek (1987) *An expert system for remote sensing.* IEEE Transactions on Geoscience and Remote Sensing, 25, pp. 249-259.

Hinton, J. (1996) *GIS and Remote Sensing Integration for Environmental Applications.* International Journal of Geographical Information Systems 10(7), pp. 877-890.

Hutchinson, C. (1982) *Techniques for combining Landsat and ancillary data for digital classification improvement.* Photogrammetric Engineering and Remote Sensing, 48, pp. 723-740.

ISO (1996) *Homepage of the Technical Committee 211 (Geoinformation) of the International Standards Organization*, http://www.statkart.no/isotc211/.

Jackson, M. and D. Mason (1986) *The development of integrated geo-information systems.* International Journal of Remote Sensing, 7, pp. 723-740.

Kwok, R., C. Curlander, and S. Pang (1987) *Rectification of Terrain Induced Distortions in Radar Imagery.* Photogrammetric Engineering and Remote Sensing, 53, pp. 507-513.

Lees, B. and K. Ritman (1991) *Decision Tree and Rule Induction Approach to the Integration of Remotely-Sensed and GIS Data in Mapping Vegetation in Disturbed or Hilly Environments.* Environmental Management, 15, pp. 823-831.

Matsuyama, T. (1987) *Knowledge-based Aerial Image Understanding Systems and Expert Systems for Image Processing.* IEEE Transactions on Geoscience and Remote Sensing, 25, pp. 305-316.

McKeown, D. (1987) *The Role of AI in the Integration of Remotely Sensed Data with GIS.* IEEE- GRS.

Michalak, W. (1993) *GIS in Land Use Change Analysis: Integration of Remotely Sensed Data into GIS.* Applied Geography 13(1), pp. 28-45.

Miller, D. (1994) *Coupling Process-Based Vegetation Models to GIS and Knowledge-Based Systems With Reference to Vegetation-Change*. In Worboys, M. (Ed.) Innovations in GIS, 1, London: Taylor & Francis, pp. 241-250.

Nale, D. and J. Thurgood (1996) *Integrating Space-based Imagery With GIS Landbase Datasets*. GIS World 9(6), pp. 58-60.

OGC (1996a) *The OpenGIS Abstract Specification: an Object Model for Interoperable Geoprocessing*, Revision 1. OpenGIS Project Document Number 96-015R1. Open GIS Consortium: Wayland, MA.

OGC (1996b) *The OpenGIS Guide: Introduction to Interoperable Geoprocessing*, OpenGIS TC Document Number 96-001. Open GIS Consortium: Wayland, MA.

Pyka, K. and K. Steinnocher (1994) *Auswahl eines optimalen Datensatzes für die multispektrale Klassifikation unter Einbeziehung von Texturmerkmalen*. Zeitschrift für Photogrammetrie und Fernerkundeung, 62(4), pp. 116-122.

Rasmussen, J. (1986) *Information Processing and Human-Machine Interaction: an approach to cognitive engineering*. New York: North Holland.

Rumbaugh, J., M. Blaha, W. Prmerlani, F. Eddy and W. Lorensen (1991) *Object-Oriented Modeling and Design*. Englewood Cliffs, NJ: Prentice-Hall.

Shi, W. and M. Ehlers (1996) *Determining uncertainties and their propagation in dynamic change detection based on classified remotely-sensed images*, International Jornal of Remote Sensing, Vo. 17, No. 14, pp. 2129-2741.

Shi, W., (1994), *Modelling Positional and Thematic Uncertainty in Integration of GIS and Remote Sensing*. PhD Dissertation, University of Osnabrück-Vechta, ITC Publication 22, Enschede, NL.

Smith, B. (1989) *Integration of Remote Sensing with GIS for the Purpose of Urban Analysis*. Geography 506: Introduction to GIS.

Star, J., J. Estes and F. Davis (1991) *Improved integration of remote sensing and geographic information systems: a background to NCGIA Initiative 12*. Photogrammetric Engineering and Remote Sensing, 57, pp. 643-645.

Trotter, C. (1991) *Remotely-sensed data as an information source for geographical information systems in natural resource management: a review*. International Journal of Geographical Information Systems, 5, pp. 225-239.

Trotter, C. and J. Dymond (1993) *The acquisition and application of digital elevation data using GIS and remote sensing*. In Proceedings of the International Geoscience and Remote Sensing Symposium, (IEEE: Tokyo), pp. 1713-1716.

Wilkinson, G. (1993) *The generalization of satellite-derived raster thematic maps for GIS input*. Geo-Informations-Systeme 6(5), pp. 24-29.

Wilkinson, G. (1996) *A Review of current Issues in the integration of GIS and remote sensing data*. International Journal of Geographical Informations Systems, 10 (1), pp. 85-101.

Wilkinson, G., I. Kanellopoulos, N. McCormick, K. Fullerton and J. Mégier (1995) *Fractals in Geoscience and Remote Sensing*. Image Understanding Research Series 1, Luxembourg: European Community, Report EUR 16092 EN.

Woodsford, P. (1994) *Integration of Remote Sensing and GIS*. International Society for Photogrammetry and Remote Sensing (ISPRS) Commission II Symposium, Vol. 30 Part (2), pp. 383-390.

Semanitic Modeling for the Acquisition of Topografic
Information from Images and Maps
ed. by W. Förstner and L. Plümer

Methods of Data Base Interpretation - Applied to Model Generalization from Large to Medium Scale

Karl-Heinrich Anders and Monika Sester

Institute of Photogrammetry
Stuttgart University
{kh.anders,monika.sester}@ifp.uni-stuttgart.de

Abstract

ABSTRACT:

Today huge quantities of digital data sets are being acquired - usually for different purposes. One important problem is the possible use of the data for new applications - originally not forseen in the initial modelling process. In order to allow this, tools and techniques are necessary to derive implicit information from the explicitly stored data. Existing GIS-products offer only very few facilities to allow such a functionality - usually it is realized based on an interactive use of various analysis tools.

An automation of such functionalities can be achieved by different techniques, namely Data Mining, Machine Learning or methods from Computational Geometry. The extraction of the implicit information also bases on an explicit semantic model of the objects affected by the process. Object oriented programming techniques offer a straightforward and elegant representation of this.

The paper focusses on the possible applications of data base interpretation and formulates the need for the development of techniques that allow such a functionality.

1 Introduction

Growth of the available quantities of digital spatial data naturally leads to the question of how these different data sets can be integrated and maintained in a unifying way. Usually different data bases are established for different purposes: as an example, national mapping agencies derive topographic information, urban planning institutions acquire data for city planning, traffic management companies need detailed road information for the autonomous navigation of cars - all of them operating on the same physical reality, namely the earth's surface, and sometimes even on the same type of objects. The problem is that usually the underlying application determines both the data model and also which data is acquired at which level of detail. Thus although a huge amount of spatial data is already available - the question remains whether the data can also be used beyond their original purpose. This question is closely linked to the problem of derivation of information, which is only implicitly given in the data set. Imagine a data set representing topographic objects as visualized in Figure 1. The data - taken from the german ALK[1] - describe legal aspects.

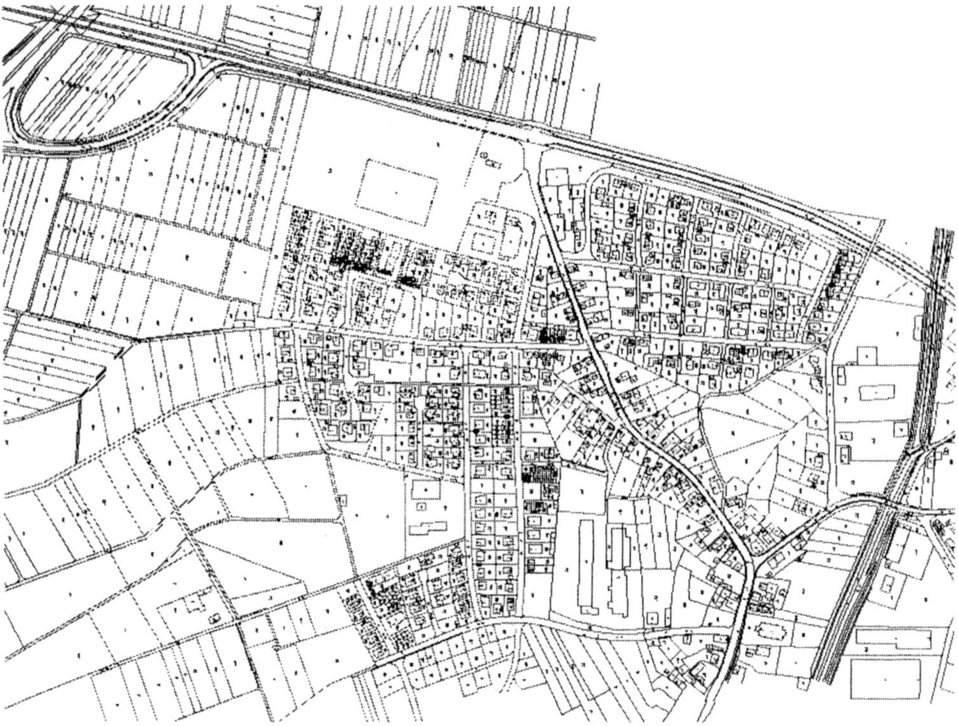

Figure 1: Digital cadastral map

Although in the data set primarily only lines and text symbols are stored, we as humans can easily read more information from it: we can identify individual parcels and road crossings, distinguish city area from rural region, detect the neighborhood between roads and built-up areas, ... Thus a human viewer can not only see the explicit information, but also the implicit one. The paper deals with the problem to automate this functionality of exploiting also implicit information in a spatial data base and making it explicit for different application areas.

The paper is organized as follows: after an introductory overview of possible applications for the derivation of implicit information from spatial data sets, the

[1] Automatische Liegenschaftskarte: Digital Cadastral Map

technological background for this is described, namely machine learning and data mining techniques, as well as methods from computational geometry. Section 4 describes one application in detail. Finally a discussion and an overview of possible extensions of the approach is given.

2 Motivation, Applications, and Related Work

Looking for implicit information in data bases is a discipline called Data mining - up to now mainly applied to descriptive data sets. An example is the prediction of the audience of BBC-customers based on audience figures like type of program, time of showing, nature of competing programs (ISL 1996). The application of this technique to spatial data bases has been addressed only recently. Here however there is a huge potential of possible applications:

Models for Image Interpretation:

Automatic image interpretation heavily relies on models to guide the interpretation process. These models are usually designed by hand and then applied to the image data. On the other hand huge collections of object models are already available in terms of objects stored in digital databases like CAD-systems or GIS. This information however cannot be used directly, because usually not all the necessary information for image interpretation is stored explicitly. For example Plietker (1994) uses ATKIS-data[2] to find roads in aerial orthophotos. In order to transfer the ATKIS-model to the image interpretation process the data has to be enriched by general knowledge about it, e.g. the fact that roads are captured with their middle axis and with a predefined width. Haala and Anders (1996) give another example of the use of individual object instances from a GIS for image interpretation: the extraction of buildings from stereo data using 2D-building groundplans. As only 2D information is available, hypothesis about the 3D-form and shape of the buildings have to be inferred from the data set.

Data interpretation can also be used to derive an intensional, generic description from the extensional set of individual object instances in the data base. Thus a generic object model can be achieved. This general model comprises not only object attributes but also relations between objects. The latter being very important since they specify the context (cf. (Sester 1995)).

Multiscale Representation / Model Generalization:

Each data acquisition process starts with modeling the data to be captured. The model is determined by the application, and thus describes that part of the reality, which is "seen" from the system. In order to make use of data also for other applications than they were originally acquired for, a transformation of the data has to be performed. This presumes to have rules for the transition between different data models. The more different the models are, the more

[2] Authoritative Topographic Cartographic Information System

complex are the relations between the data sets and thus is the derivation of the rules or transformations. This is even more true for the case when the data stems from different resolutions of detail (in terms of geometry, semantics and time). E.g. ATKIS specifies built-up areas - in order to derive them from ALK (which describes legal boundaries graphically - including buildings) a relation of scales 1:1500 with 1:25000 has to be found. This can be achieved using grouping operations, which base on spatial neighborhood. See section 4 for an implementation of such an aggregation method.

van Oosterom (1995) describes a so-called reactive data structure - a data structure representing levels of detail. His GAP (generalized area partitioning)-tree is successfully applied to perform a generalization on-the-fly - primarily for visualization purposes. His aggregation rule is defined a priori and bases exclusively on object adjacency.

The approach given by Molenaar (1996) specifies a semantic model of the aggregation hierarchies in advance. The aggregation itself then is based on spatial adjaceny.

Use of data sets for different applications/map matching:
When the scales of the different applications are similar, map-matching techniques can be applied. This process allows for a transformation of one data set into the other. Usually matching is based on geometric similarity (e.g. position, form, ...); but also semantic similarity is of great use. An example for such a matching is the link between GDF[3] road data and ATKIS roads. Although the data sets reflect the same physical reality (namely roads) and are based on the same resolution (1:25000), due to slightly different underlying data models different object instances are acquired. The approach by Walter and Fritsch (1995) uses relational matching to find the correspondences between the data sets.

Another application is the integration of application dependent information into existing data sets, e.g. the integration of environmental information into an official topographic map.

Derivation of Meta-Data:
In order to allow for a more flexible, open and independent use of digital data, so-called meta-information is of great importance. Meta-data specifies general knowledge about the data (e.g. object types, extremal values, range of values, ...). The problem with meta-data is that it is usually not stored explicitely with the data, but mostly is only known by its intelligent user. This information can be derived from actual object instances of the data set, by spatial interpretation techniques.

Extraction of High Level Structure:
City planners need information about typical structures in a city. In order to

[3]Geographic Data File: standard for road data in Europe

derive this knowledge, an analysis and statistical evaluation of the facts and the relations between individual objects has to be performed. This process can be automated based on a given dataset by spatial data interpretation. This presumes to have a model of what is to be found. An example is a data set consisting of roads; if road crossings have to be found, a model for a crossing has to be available. This interpretation task can directly be compared to image interpretation, which also needs models in order to find objects in an array of gray values.

Potentially also relations are interesting, which are not known or clear in advance, and kind-of hidden in the data. Therefore techniques are necessary to automatically derive dependencies, relations and statistical information from the data. For example a spatial and statistical analysis of data in a city can reveal dependencies of certain buiding types and special regions, etc. Another example are the requirements for regional planning purposes, where information about the population density is very important. This should be derived relating to special regions, along roads, ... Koperski and Han (1995) give an example of spatial data mining where they analyze the adjacency of canadian cities and other topographic objects. The result is that huge cities tend to be close to the sea.

All these examples demonstrate how the analysis of spatial data bases can be used to extend the application spectrum of a digital data set.
In order to exploit the implicit information and make it explicit, several techniques are necessary. Basically they can be seen as generalization operators: the specific information given in the examples has to be generalized to an intensional, generic description valid for a broader range of objects and applications. Thus first the object description itself has to be generalized by adequate methods; furthermore also the spatial relations among them have to be focused on. Such spatial relations need information about neighborhood. The notion of neighborhood can be used in different context:

- ▷ physical connectedness, adjacency (e.g. common border)
- ▷ closeness, nearness (distance less than a given threshold)
- ▷ thematic neighborhood (similarity)

Thus a fundamental prerequisite for a spatial analysis operator in a GIS are neighborhood or grouping operations based on - application dependent - neighborhood.

3 Techniques for spatial data interpretation

Data Mining

Data Mining tools aim to extract high value information from low level data. The background being the fact that stored raw data does not necessarily correspond to stored knowledge. The tools for the extraction of knowledge from the data can be classified into interactive and automatic. Graphical analysis tools fall into the first class, e.g. histograms, distributions or plots for 2D-distributions. In this case it is up to the user to select certain views of the data and analyze them. There are however also automatic techniques, where the system is able to find the underlying structure in a data set. Among them are model building technologies like classification, segmentation, rule induction, neuronal networks, regression modelling, rough sets, ... The applications lie primarily in the domain of the analysis of attribute data, and range from financial risk assessment to electicity demand prediction (ISL 1996). The extension to the interpretation of spatial data bases has evolved only recently. The principal problem being the specification of spatial operators.

Machine Learning

Machine Learning (ML) is a discipline from Artificial Intelligence dealing with the problem of inducing information from a set of examples to achieve general information (see e.g. (Michalski et al. 1984)). In contrast to deduction, however, induction never guarantees that the result is true for all examples which can occur later on - such inferred rules and statements can only be falsified, not verified. Nevertheless induction is an important principle which also governs the human knowledge acquisition process. In order to transfer this to a machine, there is a abundance of approaches, ranging from simple parametric learning to unsupervised learning or conceptual clustering. Also Neural Networks (NN) fall into this category; the main difference between ML and NN being the fact that the result of a ML-algorithm somehow allows the verification - since it is mostly given in terms a set of rules, category descriptions or decision trees. In the case of NN however the rule is hidden in an array of nodes and their relating links with weights.

A very popular ML-approach is ID3 by Quinlan (1984), which belongs to the class of "learning from examples". Input is a set of classified examples, given in terms of attribute value lists. ID3 then derives a decision tree that allows for a classification of unclassified examples based on the attributes. The underlying principle stems from information theory: the attribute with the highest information content with respect to the given examples, is selected first. Thus the aim is to choose that attribute, that arrives at a best distinction of the given classes. The charm of this approach is that it can be applied to any data set given in the assumed structure, namely attribute value lists. The drawback is that only discrete attributes can be used; each attribute value is viewed as being distinctive. In order to be able to treat continuous values as well, extensions of ID3 have been proposed to first cluster the attributes into discrete groups before they are introduced into the program ((Quinlan 1993)).

Sester (1995) describes the use of this program for the derivation of models for image interpretation from given visual examples. The learning strategy essentially relies on a close interaction between the setup of classification trees for concepts and relations on one hand and the immediate application these concepts for interpretation. Automatic programming techniques are used in an object oriented interpreter system, supporting inheritance between concepts, smooth transition between learning and interpretation and a transparent representation of the learned concepts. This results in a fast and efficient knowledge acquisition of complex scene models. Both, concepts and semantic relations, are learned with the same technique generalizing the given examples. They lead to a specialization and a containment hierarchy as well as in an association network. In this way the underlying structure of a data set can be revealed in an interactive process between a teacher specifying examples and a learning system deriving the corresponding parameters for it. Thus the relations in a data set can be modeled according to a given application.

Computational Geometry

In order to reveal spatial neighborhood between objects, a triangulation of the points an object is composed of can be performed. Bundy et al. (1995) describe the so-called SDS (Simplicial Data Structure) to represent geometric objects in a data base. The SDS is a explicit triangulation that provides a rich data structure to describe topological and proximity relations between spatial objects in a map (Jones et al. 1996). Two objects are considered as neighbors, if they are connected via a common edge in the triangulation network. They applied this data structure especially to perform (cartographic) generalization procedures, like merging or displacement.

Such a data structure can be exploited to a certain degree. An extension is to evaluate also the length of the vertices between different objects (and thus rating different degrees of neighborhood). Also the type of objects connected can be important for further processes.

After different clusters of objects have been detected (e.g. minimal spanning trees), a convex hull around these objects can be used to derive the new aggregated object.

4 Generalization from Buildings to Settlement Areas

4.1 Basic assumptions and available data

In Germany the ATKIS data base is established in three levels of detail, namely 1:25 000, 1:200 000, 1:1000 000. This data set comprises topographic objects of seven different classes (e.g. settlement, traffic, vegetation, ...). On the other hand, cadastral information in scales 1:1500 ... 1:2000 is acquired for the ALK. In ALK legal boundaries, as well as useage or type of objects are specified. In order to facilitate and automate the time consuming update process, the idea arises to derive ATKIS data from ALK. In this way only the more detailed data set (ALK) would have to be updated manually - whilst the other can be deduced from it. Currently the update is done separately for each data set; ATKIS is acquired (in Baden-Württemberg) based on a manual interpretation of orthophotos, while ALK stems from direct geodetic measurement. Using ALK for ATKIS update involves generalization or aggregation operations, respectively. First of all the links between the data sets have to be established.

In the following, an approach will be demonstrated showing the derivation of built-up areas in ATKIS from ALK data. A so-called object-catalogue of ATKIS specifies, how the objects are to be created and captured. The following "rule" describes how built-up areas (object id 2111) have to be acquired (*ATKIS-Objektartenkatalog* n.d.).

> "Built-up areas are areas that are exclusively or mainly used for living. Besides buildings, there are also objects serving for the provision of needs of the whole area like small workshops, organizations of clerical, cultural, social, or sanitary purpose.
>
> The boundaries between a built-up area and neighboring areas is given by the boundaries of the parcels of land the buildings are standing on.
>
> Only areas greater than 10 hectars have to be captured; if the area is smaller, it has to be acquired only in special cases.
>
> The geometric object type is a point (for areas less than 10 hectars) or an area for those greater than 10 hectars."

The first step is to implement such a rule. The specification above reveals a link between the data sets, namely the fact that built-up areas consist of adjoining residential areas. Residential areas are parcels of land comprising building objects like houses, garages, ... This however is information which is stored (explicitly and implicitly) in ALK. As the rule implies to aggregate neighboring building-objects to a settlement area, the question is how to define neighborhood in this application. First of all, in order to come from ALK to ATKIS, the underlying data models have to be examined in detail in order to make them explicit and thus usable. This will be revealed in the following semantic modeling process.

4.2 Semantic modeling, data structures, and implementation

The digital cadastral map (ALK) is provided by the landsurvey administration. Possible formats the user can get the data are SICAD/open data exchange format (SQD) or DXF. In these formats these data sets contain only geometric features and text elements. These elements are points, lines, areas and text features. SICAD/open additionally provides the feature parcel number. Thus the data is a mere graphical representation (cf. Figure 1).

In order to describe how spatial objects are represented in a digital cadastral map, we use *semantic modeling* as a conceptual method to analyse how a human operator detects such kinds of objects. Figure 2 shows the semantic model which we use to describe parcels of land and residential areas.

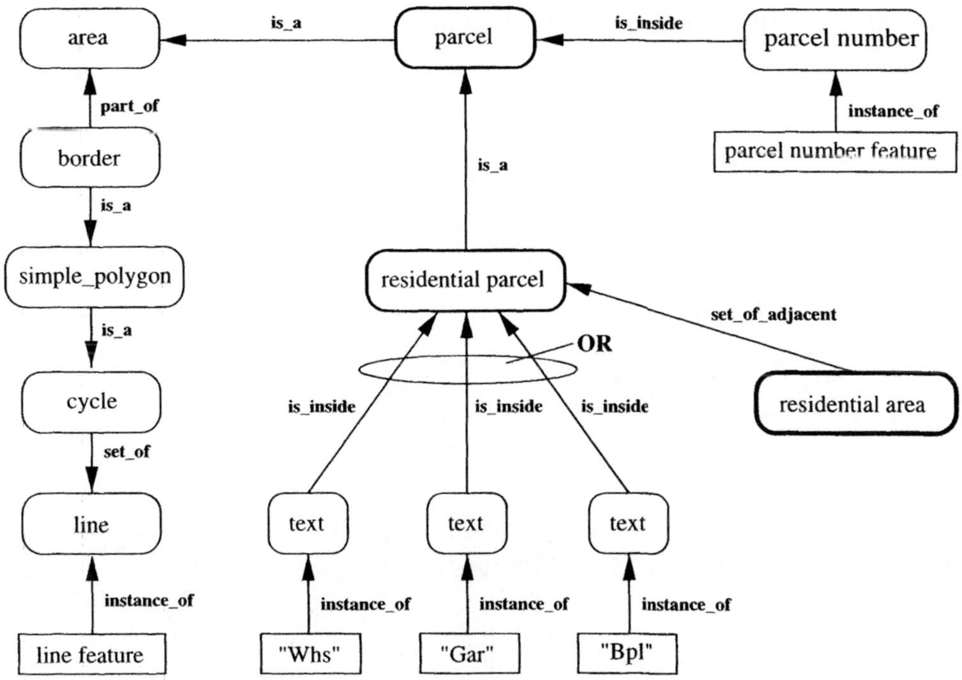

Figure 2: Semantic model of parcels and residential areas

To represent the spatial objects in the computer we use an object oriented data model presented in (Anders and Fritsch 1996), (Fritsch and Anders 1996). *Spatial objects* are aggregates of *thematic objects* which describe the semantics of an object, and *geometric objects* which represent the spatial position and form of an object. This structure directly allows to compute the topological relation *adjacent_to* among the features, because the incidence relations for the topological objects point, line and area are stored explicitly in our data model. As every line object "knows" to which area it belongs, the adjacency between areas can be computed. The whole implementation bases on an object oriented database in conjunction with an object oriented programming language. The advantage is that we are able to represent semantic relations like *is_a* and *part_of* directly in our data model.

The first problem is that in the given representation of the digital cadastral map parcels are not stored as area features. Only the borderlines of the parcels are stored as line features. The buildings are directly stored as area objects - but this is only true for most of the cases. Therefore the first step in our approach is to compute all possible areas based on the given list of line features. An area is represented through a *simple polygon* which is a cycle of line features in the net of lines.

In the next step the relation between geometric objects parcels and the corresponding parcel number features have to be established. This can be achieved by geometric calculations of a point-in-polygon test between all possible combinations of parcels and numbers. This results in the following figure 3.

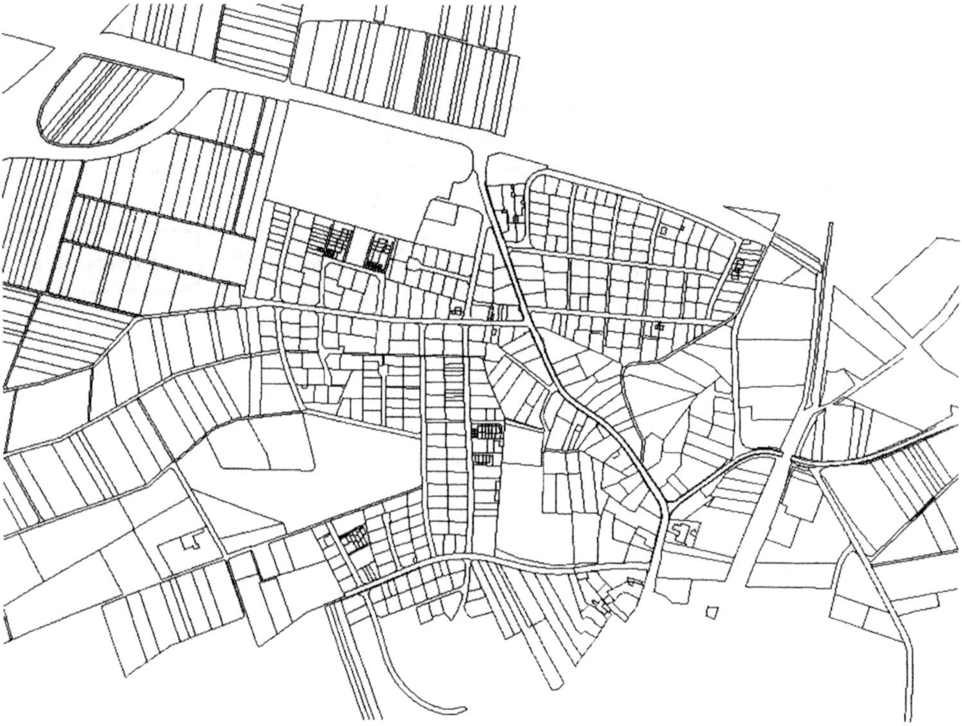

Figure 3: Extracted parcels

Then we have to choose all that parcels which probably have a residential use. Therefore we take all text features which are instances of *'Whs'*, *'Gar'*, *'Bpl'* (building, garage, building plot) and again make a point-in-polygon test for all possible combinations to find the relations to the corresponding areas.

In the last step we have to group all computed parcels with residential use to create *residential area* objects. The semantic rule for grouping of these areas is the adjacency principle (see right hand side of Figure 2). I.e. all residential areas which are connected via a common border are merged. This can easily be performed using the topological relations between the areas which is directly extractable from our data model.

Figure 4 gives the result of the grouping and shows the built-up areas in that region.

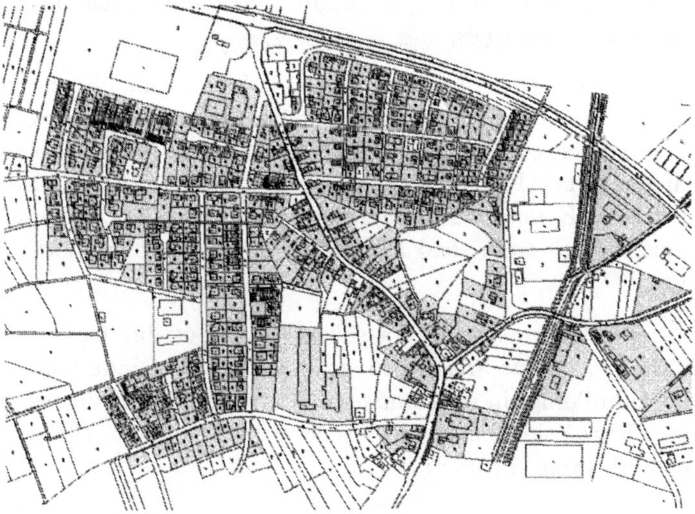

Figure 4: Built-up areas generated from ALK data

This result was achieved only based on the assumption of a spatial aggregation of adjoining residential objects. Comparing these ATKIS-objects with original ATKIS-data reveals the following differences (see Figure 5):

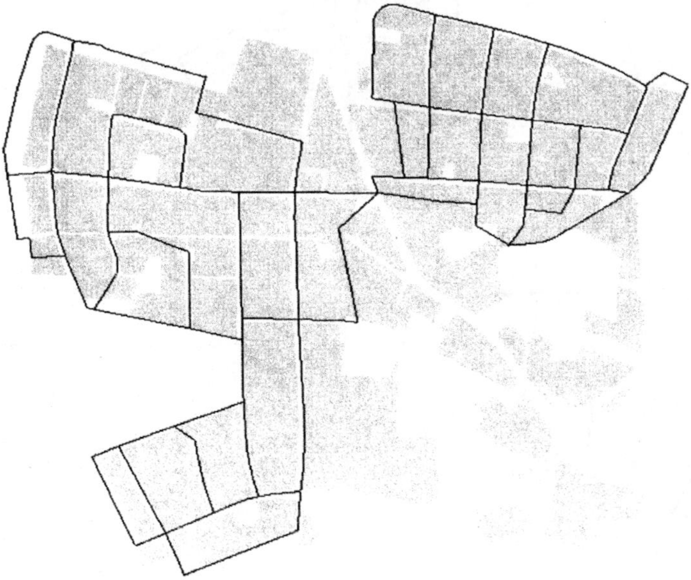

Figure 5: Overlay of original ATKIS built-up areas (black lines) and automatically derived areas (shaded areas)

▷ The main source of the difference are: possibly different acquisition dates, different acquisition rules, ...

▷ The boundaries of the areas do not exactly correspond: this is due to the fact that in ATKIS the roads are stored as line objects only by their middle axis - the adjacent objects then meet at that axis; in our case as we take the road model from ALK (which is area based), the built-up areas consequently border the ALK roads. An extension would be to also generalize the roads (find middle axis from road areas) and then generalize both data types in common.

▷ In the current approach only a subset of all possible building objects have been modeled. Thus areas with mixed usage (like industrial buildings and residential buildings in the same parcel) are included in our result - in ATKIS however they should belong to a new class. Therefore the model has been modified reflecting this fact:

> If a parcel contains an industrial building then it is no residential area.

Figure 6 shows the improved result. Direct comparison with figure 4 shows that less regions are classified as built-up areas now.

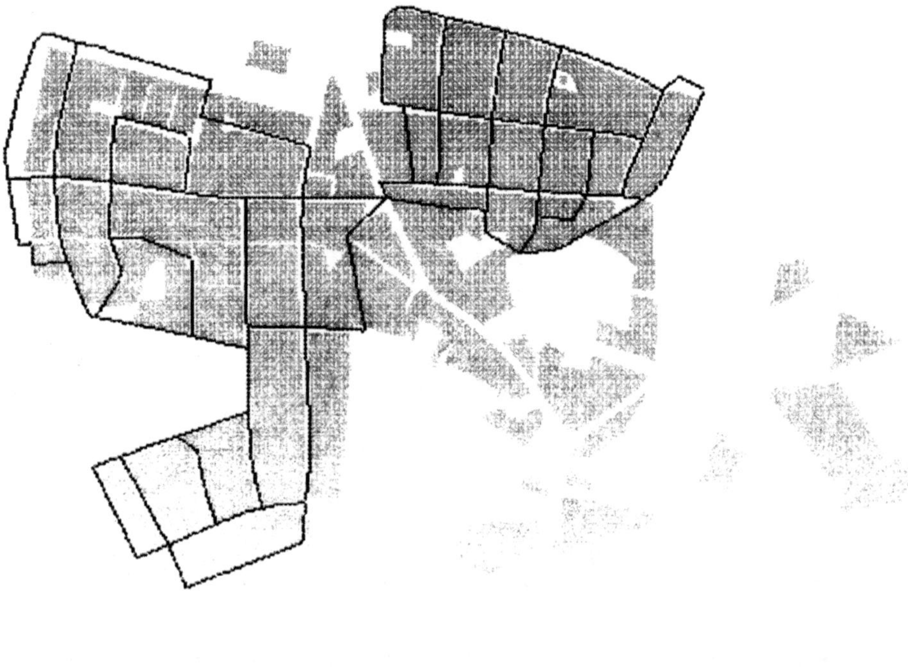

Figure 6: Built-up areas generated from ALK data excluding industrial buildings

4.3 Extensions

The result of the previous steps shows, which aggregation level is achievable based on an initial semantic modeling of the knowledge implicitely given in the data and explicitely given in terms of an - rather abstract - object description catalogue. In order to compensate for possible exceptions to this, different strategies have to be applied depending on the type of relation to be modeled.

If the relations are mainly geometric, a triangulation structure can be used to determine the neighborhood. This process can be followed by a spatial clustering operation, e.g. minimal spanning trees (Regnauld 1996). In the same way e.g. higher level structures of ATKIS can be derived by application of further aggregation procedures. E.g. the boundary of a city (ATKIS object id 2101) is described in the object catalogue as comprising all the built-up and industrial areas, and also including the traffic and vegetation areas. In this case the aggregation cannot base on the adjacency principle alone, but has also to consider possible gaps in between. This can be solved with a triangulation of the settlement objects, followed by a spatial clustering. The hull around these clustered objects then represents the boundary of a city.

If however the relations between the objects are not known explicitely, machine learning techniques can be applied. This approach should be used, when examples for such relations in the data set exist - when however their reasons or explanations are not known. Then the input to an ML-program is a set of examples specifying the data sharing the relations, together with a corresponding classification (the type or name of a relation). The ML-system is then able to reveal the underlying links and describe them in terms of semantic concepts or rules.

5 Conclusion and Outlook

Data base interpretation techniques are one tool to make the data re-use possible and also allow for a utilization of the data beyond their original purpose. Three basic methods have been presented. An application verified the usability of such techniques in the domain of data abstraction (model generalization).

The example showed the usefulness of such techniques for the update of digital databases like ATKIS with the help of existing ones of a higher level of detail, namely ALK. The result shows that the fusion of two databases can very efficiently be implemented based on a semantic data model and on spatial proximity operators. In the example the transfer of the semantic rules for object generation into an aggregation scheme is straightforward. Already the implementation of these first steps reveal the great potential of such techniques for data re-use. In other cases however, these links might not be that direct. Then the above described learning techniques have to be applied. This can e.g. be accomplished in a way that first a very general rule is applied - the exceptions to this rule are then incrementally acquired with the help of learning techniques.

The scope of the paper thus was to motivate the usefulness and the need for techniques of spatial data base interpretation. In times when more and more spatial data bases are build up indepentently to (enormously) costs, there is a strong need for having tools and methods to use them in a unifying way. It is obvious that no general purpose data base can ever be generated. Also, it is getting more and more clear that each geographic process has its own level of detail where it is best understood - and at this level it should be monitored and managed. Thus there is a basic need for techniques to analyze spatial data in an automatic way.

References

Anders K.-H., D. Fritsch (1996) *Automatic interpretation of digital maps for data revision*, International Archives of Photogrammetry and Remote Sensing, Vol. 31/4, ISPRS, Vienna, Austria, pp. 90–94.

ATKIS-Objektartenkatalog (n.d.) Technical report, Landesvermessungsamt Baden-Württemberg, Stuttgart.

Bundy G., C. Jones, E. Furse (1995) *Holistic generalization of large-scale cartographic data*, in Müller et al. (1995), pp. 106–119.

Fritsch D., K.-H. Anders (1996) *Objektorientierte konzepte in geo-informationssystemen*, GIS - Geo-Informations-Systeme, Vol. 9, No. 2, pp. 2–14.

Haala N., K.-H. Anders (1996) *Fusion of 2d-gis and image data for 3d building reconstruction*, International Archives of Photogrammetry and Remote Sensing, Vol. 31/3, ISPRS, Vienna, Austria, pp. 285–290.

ISL (1996) Product description demo of data mining tool "clementine".

Jones C., D. Kidner, L. Luo, G. Bundy, J. Ware (1996) *Database design for a multi-scale spatial information system*, International Journal of Geographical Information Systems, Vol. 10, No. 8, pp. 901–920.

Koperski K., J. Han (1995) *Discovery of spatial association rules in geographic information databases*, Advances in Spatial Databases '95, p. 47.

Michalski R., J. Carbonell, T. Mitchell (1984) *Machine Learning - An Artificial Intelligence Approach*, Springer-Verlag, Berlin.

Molenaar M. (1996) *The role of topologic and hierarchical spatial object models in database generalization*, in M. Molenaar (ed.), Methods for the Generalization of Geo-Databases, number 43, Netherlands Geodetic Commission, Delft, The Netherlands, pp. 13–35.

Müller J.-C., J.-P. Lagrange, R. Weibel (eds) (1995) *GIS and Generalization - Methodology and Practice*, Taylor & Francis.

Plietker B. (1994) *Semiautomatic revision of street objects in atkis database dlm 25/1*, International Archives of Photogrammetry and Remote Sensing, ISPRS Commission IV, Atlanta, USA, pp. 311–317.

Quinlan J. (1984) *Learning efficient classification procedures and their application to chess end games*, in R. Michalski, J. Carbonell, T. Mitchell (eds), Machine Learning, Springer-Verlag, Berlin, pp. 463–482.

Quinlan J. R. (1993) *C4.5: Programs for Machine Learning*, Morgan Kaufmann, San Mateo, California.

Regnauld N. (1996) *Recognition of building clusters for generalization*, in M. Kraak, M. Molenaar (eds), Advances in GIS Research, Proc. of 7th Int. Symposium on Spatial Data Handling (SDH), Vol. 2, Faculty of Geod. Engineering, Delft, The Netherlands, p. Session 4B.

Sester M. (1995) *Lernen struktureller Modelle für die Bildanalyse*, Vol. C441, Deutsche Geodätische Kommission, München.

van Oosterom P. (1995) *The GAP-tree, an approach to 'on-the-fly' map generalization of an area partitioning*, in Müller et al. (1995), pp. 120–132.

Walter V., D. Fritsch (1995) *Matching techniques for road network data in different data models*, in J. Soliman, D. Roller (eds), 28th International Symposium on Automotive Technology and Automation, Automotive Automation Limited, Croydon, England, pp. 633–640.

Roads

Semanitic Modeling for the Acquisition of Topografic
Information from Images and Maps
ed. by W. Förstner and L. Plümer

A Semantic Road Model for Aerial Image Interpretation

Marlies de Gunst[1] and George Vosselman[2]

[1]Netherlands Cadastre and
Public Registers
P.O. Box 9046, 7300 GH Apeldoorn
The Netherlands
Phone: +31-55-5285206
Fax: +31-55-3557931
M.E.deGunst@ap.KADASTER.nl.net

[2]Faculty of Geodetic Engineering
Delft University of Technology
Thijsseweg 11, 2629 JA Delft
The Netherlands
Phone: +31-15-2781741
Fax: +31-15-2782745
G.Vosselman@geo.TUDelft.nl

Abstract

This paper focuses on concepts to use knowledge-based techniques for interpretation of road networks in aerial images. Emphasis is on the contents of the semantic road model, which yields a knowledge base for a certain application. Image understanding, easily performed by human operators, is the bottle-neck in automatizing photogrammetry. The complex contents of aerial images requires recognition of meaningful parts of the road network instead of linear structures. The Dutch standards for road construction are utilized to define properties of specialized road types. The knowledge base build is part of a framework for a complete road network model, covering multiple scales of imagery. Besides properties of objects to be extracted, it contains also spatial, part-whole and specialization-generalization relationships. By means of an example the concepts are concretized for extraction of road sides and road markings at large scale aerial images.

1. Introduction

In the photogrammetric practice there is a need to automatize acquisition of topographic information from aerial photographs. In particular interpretation tasks are very labour-intensive and time-consuming tasks in the mapping process. However, especially tasks involving interpretation capabilities of human operators are hard to automate. This in contrary to the fact that even unexperienced people can immediately recognize for example most roads and houses in aerial images. Nevertheless, nobody can tell exactly how they did it. People unconsciously rely on knowledge about properties of topographic objects and their appearance in the aerial image. In order to perform this task by computer, this knowledge should be formulated exactly, together with techniques to measure them in the image.

The wide variety of objects and properties requires the use of knowledge-based concepts from computer vision (Draper et al., 1989), including explicit representation of knowledge about objects in the scene. However, building a formal model that includes all relevant knowledge about objects will be a difficult task. Firstly, properties of objects need to be collected from various sources and subsequently be evaluated for their relevance. Secondly, the knowledge representation formalism is required to organize and represent a wide variety of properties, which belong to objects themselves or to the interrelationships between objects. Because of the complexity to build a knowledge base which includes complete descriptions of all topographic objects and their interrelationships, the interpretation strategy in this paper is mainly designed for the extraction of one class of topographic object: road networks. Roads are a means of communication between different locations. Since they express the principle structure of an area, they can form a logical framework to search other topographic objects. Even though only roads are considered in this paper, the design of the knowledge base is required to allow easy extension for other topographic objects.

Because of the complex characteristics of aerial images, semantic information is necessary to identify meaningful individual objects, rather than syntactic information to extract structures. In this paper especially the inclusion of semantic object-specific knowledge and the organisation of this knowledge in the road model will be emphasized. For this purpose the standards for road construction (Rijkswaterstaat, 1975) are utilized. Our multi-scale semantic road model is presented in section 3. Next, in section 4 the concepts of the interpretation strategy, which emerges from the road model, are described. Finally, in section 5 the concepts are concretized for the detection of road sides and road markings in large scale aerial photographs. But first, in section 2, road models in previous work are reviewed.

2. Road Models in Previous Work

A large number of publications about road extraction appeared in the past 20 years. In early work road extraction is primarily approached as a problem of linear feature detection and grouping of lines to form a network (e.g. Bajcsy/Tavakoli (1976), Fischler et al. (1981), Groch (1982)). General geometrical properties, concerning width and straightness, and radiometric statistics, based on contrast and homogeneity of intensity are used to select those lines which are probably a road. Because these properties are too simple to distinguish for example roads from railways, roads cannot be recognized unambiguously.

As a result, knowledge-based systems, which offer the possibility to model object-specific knowledge in an explicit way, are applied for interpretation of aerial images in general and road extraction in particular (e.g. Matsuyama and Hwang (1985), Cleynenbreugel et al. (1990), Garnesson et al. (1990)). It is generally agreed that the object model must be able to include, besides properties of topographic objects, also relationships between objects which represent context, specialized classes and component parts. Only Matsuyama and Hwang (1985) use all three, spatial, specialization-generalization and part-whole relationships, in their road model. Cleynenbreugel et al. (1990) implements context as general constraints on the collinearity, parallelism and perpendicularity of segmented line elements. The implementation of Garnesson et al. (1990) supports inclusion of semantics. They implement context as spatial relationships, which generate goals to find other objects most likely present. However, in previous work the chosen road objects and properties used for recognition remain general, e.g. road pieces and intersections, and they are not related to meaningful parts and properties of the road network. For that purpose we include the standards for road construction (Rijkswaterstaat, 1975) in our road model. These standards specify how properties differ for specialized classes of roads, like motorways or streets, depending on their function.

3. Semantic Model for Road Networks

In this work a semantic model is defined as the choice of meaningful object classes, together with object-specific properties, and the relationships between those object classes. Important for a semantic approach is the definition of object classes corresponding to meaningful parts of the road network. The representation of the road network differs considerably for different scales. Therefore it was decided to discriminate between three levels of detail in the road models: small, medium and large scale. Levels are connected by part-whole relations. Object classes at each level in their turn can be seen as generalized object classes at the top of a specialization hierarchy. For interpretation of a particular aerial image at a certain scale, a knowledge base with specialized object classes of the corresponding level of detail is build. These specialized objects are connected by spatial relationships.

First, in section 3.1 the object classes of the overall road model covering multiple scales will be described. For a large scale application these generalized objects types are elaborated in section 5 into specialized object classes, relations and their object-specific properties. For medium scale this elaboration can be found in Gunst (1996). The formalism used for representation of object class and relationships is outlined in section 3.2.

3.1 Multi-Scale Road Model

The generalized object classes of the multi-scale road model are shown in fig. 1. Examples of the appearance of the object classes in fig. 1 are given in fig. 2 as a zoom on the same situation at various scales. One should notice that the designed multi-scale road model is restricted to motorways and other through roads and does not include small (unpaved) roads in towns and villages like streets. It should also be noticed that in this paper similar road terms are used with a specific meaning, expressing at which level of detail a part of the road network is defined.

At small scales, like satellite images, road networks can be represented by lines forming a network of intersections and roads. At medium and large scales the road surface is represented in the digital map as a planar feature. Intersections at small scale usually consist of many road elements and road junctions at medium scale, which all together manage traffic flow. The rest of the medium scale road elements are part of the roads at large scale. Road elements and junctions both are a composition of three generalized object classes at large scale: traffic area, road marking and road side. They are described in detail in section 5.1.

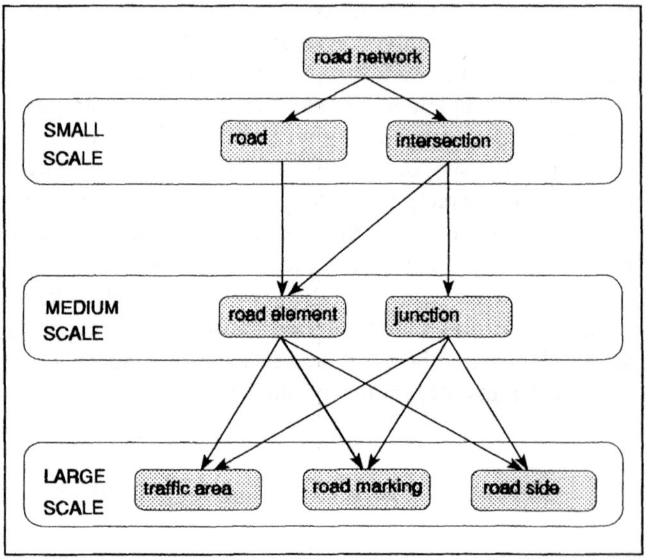

Fig. 1: Generalized object types defined at the three levels (small, medium and large scale) of the semantic model for road networks

One could imagine that there is a third dimension in fig. 1 which connects specialized object classes with generalization-specialization relationships. For example: crossing, fly-over and Y-junction are specializations of a junction at medium scale. Geometric and radiometric attributes are attached to generalized objects if they are valid for all specialized objects. Attributes attached to specialized objects serve to discriminate them from each other. Using the standards for road construction (Rijkswaterstaat, 1975), discriminating properties for recognition of object classes need to be defined. Specialization-generalization relationships are not only useful for definition of objects with more specific properties, but they also yield semantic constraints between objects. Motorways for example cross each other by fly-overs and not by level road junctions.

The position of specialized object classes with respect to each other is modelled by spatial relationships. For example: a lane line on the carriageway of a two-lane motorway runs parallel to both edge lines at a distance of 3.5 meter. The standards for road construction (Rijkswaterstaat, 1975) are used to define spatial relationships. Spatial relationships can help to determine where to expect a certain object, but also yield constraints for the presence of objects.

3.2 Knowledge Representation Formalism

The previous section shows that the knowledge which needs to be represented is enormously varied. Because declarative as well as procedural knowledge are involved, frames are the most suitable representation formalism. Frames also support object-oriented reasoning and allow easy extension for other topographic objects. The underlying semantic network reflects the discrimination between object and relational knowledge.

Because we use an object-driven fashion of reasoning, procedural knowledge should be attached to the relationship between the known and searched object type. It consists of knowledge about image processing techniques for goal-directed segmentation and requires that the relation between the searched object class and appropriate segmentation techniques is known.

The represented knowledge is divided into two types of basic representation primitives (Hartog, 1995): object definition and object relation.

The *object definitions* contain a description of geometric and radiometric properties for each object type. They are used to recognize segmented objects. Only relevant properties are specified. Properties described in the standards for road construction are valuable to use for recognition. A range of allowed values is attached to each feature, providing a condition for recognition of the object. Because this range should be defined a priori, extensive testing will be necessary to determine appropriate values.

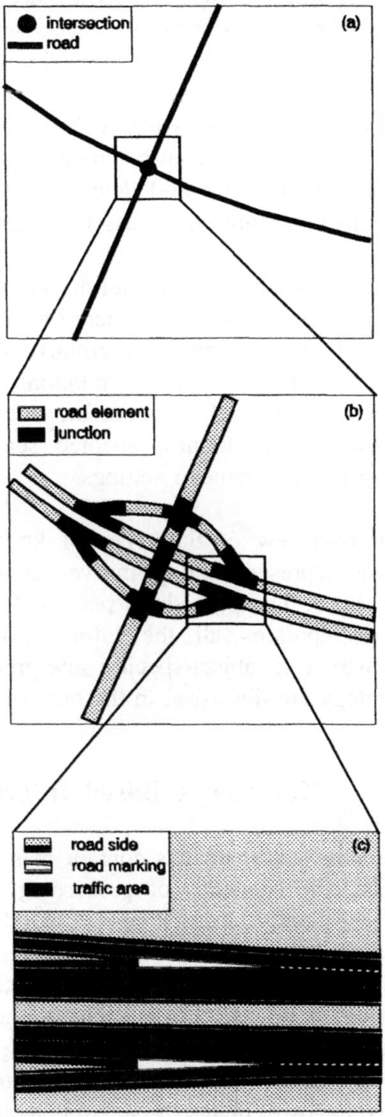

Fig. 2: Example of the same situation at various scales with corresponding objects (a) small scale, (b) medium scale, (c) large scale

The *object relations* describe spatial, specialization-generalization and part-whole relationships between object classes, in this way constructing a complex semantic network which describes both declarative and procedural knowledge. The declaration part describes the location of objects with respect to each other, while the procedural part of the knowledge specifies where to search and with which image processing technique. Object relations are used to control processing by means of attached priorities and detect inconsistencies by means of definitions of essential relationships.

Variations upon object definitions and relations, which are used for generation of alternative hypotheses, are called alternative object definitions and reprocessing relations respectively. Reprocessing relations define alternative segmentation techniques, adapted search areas or alternative parameter settings.

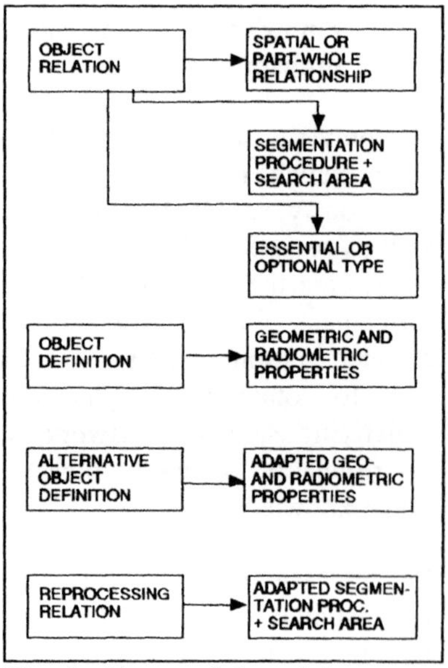

Fig. 3: Types of knowledge represented in object definitions and object relations

An overview of the types of knowledge these basic representation primitives contain is given in fig. 3. The procedural part of the object relations contains all the information needed to construct an object-specific interpretation strategy. The concepts of the interpretation strategy are discussed in the next section.

4. Knowledge-Based Interpretation Strategy

Because aerial images contain many objects with large variations within each object class, the sequence of processing steps cannot be determined reliably beforehand. Consequently, like Draper et al. (1989), the interpretation strategy is based on the classic hypothesize and test paradigm. A top-down approach generates hypotheses for the presence of objects. Most hypotheses lead to the generation of goals for segmentation of objects, comparable to the location controller of Garnesson et al. (1990). For objects with complex internal structures we need to search for parts of objects. Goals operate on a restricted region of interest in the grey-value domain. Each goal has its own image processing technique and values for its parameters, especially suitable for segmentation of a specific object type. We call it goal-directed segmentation. By generation of goals, combination of several image processing techniques is realized as required (Matsuyama, 1987).

The need of context to interpret images of complex object spaces leads to the conclusion that spatial relationships should be embedded in the interpretation strategy. Our strategy is like Hartog (1995) based on the perception that each object class is spatially related to one or more other object type. In this sense context is used to generate hypotheses for the search of new objects in spatially related to objects that are recognized previously. Recognized related objects on their turn generate hypotheses for further search of other objects that appear in the context of the detected object. Thus a chain of successive search action generation and object detection can be distinguished. As object detection might result in multiple search actions, a mechanism to handle search actions and execute them in a proper sequence is required. We have chosen for a simple and effective method using a sorted list. Each action has a predefined priority attached to it. By assigning highest priority to search actions for the object classes most likely present, the reliability of the whole interpretation will increase. We call this entire process contextual reasoning. It provides a mechanism to control generation of goals for change detection and segmentation and it embeds goal-directed segmentation.

A priori defined properties of the searched object class are used for testing of hypotheses and result in object recognition. Since errors are inevitable in image analysis, inconsistencies need to be detected and solved. For this purpose constraints in the knowledge base define if objects are required to be present in relation to other objects present. If such an object is not found, alternative hypotheses are generated, resulting in re-segmentation with adaptation of parameters, an alternative segmentation technique or an adapted search area.

Summarizing, the interpretation strategy has four processing steps (Fig. 4):

- (alternative) hypothesis generation;
- goal-directed segmentation;
- object recognition;
- inconsistency detection.

Hypothesis generation could start with bottom-up processing as applied in Cleynenbreugel et al. (1990), but we encourage the use of objects from outdated or smaller scale digital maps for this purpose. The presented interpretation strategy is also suitable for updating of road maps (Gunst, 1996). It can operate fully automatic, but it can also be used for semi-automatic road detection by providing manually extracted objects to start interpretation and/or define semi-automatic segmentation techniques in the object relations of the knowledge base.

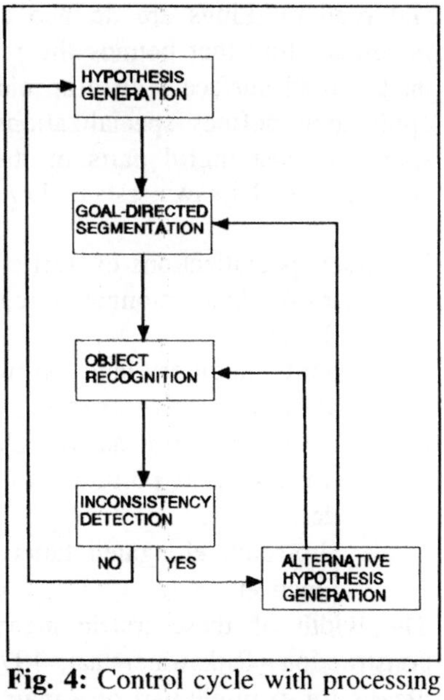

Fig. 4: Control cycle with processing steps

5. Example of Interpretation at Large Scale

In order to illustrate the concepts, an example will be given of how a knowledge base for interpretation of large scale aerial images can be build from the semantic road model. In Gunst (1996) the concepts are concretized and tested exhaustively for a case applied at medium scale.

In section 5.1 the generalized and specialized object classes are defined, which are chosen at large scale. Next, in section 5.2 an interpretation strategy for extraction of the road side and road markings is constructed. In section 5.3 the object definitions and object relations applied for extraction of one type of road marking are given. The results of extraction of all road sides and road markings are given in section 5.4.

Ranges of attributes in object definitions and parameters for goal-directed segmentation used in this example were not tested exhaustively, because the aim of this example is only to illustrate the concepts.

5.1 Choice of Object Classes

At the large scale level of the multi-scale road model (Fig. 1) road side, traffic area and road markings are defined as generalized object classes. The road side is the imaginary line that bounds the road pavement. A traffic area is defined as a region on the road surface with a special function for traffic. The knowledge base for this application defines specializations of these generalized object classes. They correspond to meaningful parts of the road network as used in the standards for road construction (Rijkswaterstaat, 1975).

The next specializations of traffic areas are defined:

- traffic lane: elongated region with parallel sides where one is supposed to drive;
- hard shoulder: lane on which only in exceptional circumstances traffic is allowed to drive or stop;
- correction strip: narrow elongated region next to the road side meant to give drivers the opportunity to correct their course in case there is no hard shoulder;
- slip-road: elongated taper lane used to join or leave the motorway at a Y-junction.

The width of these traffic areas is defined accurately in the standards for road construction (Rijkswaterstaat, 1975), but differs depending on the fact if the traffic area is for example part of a main carriageway or link road at medium scales.

Road markings are symbols painted on the road surface in order to guide, warn and control the traffic flow. There are many of them, but next are used in this section:

- triangular mark, which indicates that a slip-road branches off the main carriageway by means of a link road;

- edge line, a white unbroken line which marks the separation between the outermost lanes and the hard shoulder or correction strip;
- lane line, a broken line between two traffic lanes;
- block line, row of square blocks to indicated the separation between a traffic lane and a slip-road.

5.2 Interpretation Strategy for Extraction of Road Sides and Road Markings

An example of an interpretations strategy that can be designed at large scale is represented in fig. 5 as a semantic network. It aims to detect the road side and different types of road markings part of a specific type of road junction, a so-called Y-junction, at medium scale. It is assumed that the planar feature being the Y-junction is acquired manually, extracted previously automatically, or originates from a smaller scale map that needs to be updated. Each relationship in the semantic network corresponds to an object relation describing how to extract one object in relation to the other and the constraints. Reprocessing relationships may be defined with alternative search area, segmentation technique, or different parameters of the same segmentation technique. The priorities regulate the sequence within which actions to search certain objects are generated. In this way an interpretation strategy is constructed.

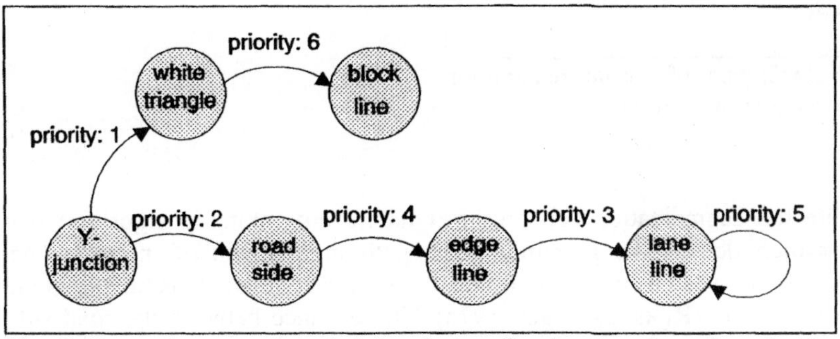

Fig. 5: Semantic network underlying frames for the detection of road sides and road markings from a medium scale Y-junction in an incomplete map. Relations correspond to hypotheses which can be generated.

5.3 Object Relations and Definitions for the Extraction of Edge Lines

For each node and relationship in the semantic network of fig. 5 the conditions for object recognition and goals for segmentation need to be described by means of object definitions and object relations. As an example of how these concepts are filled in, the used object definition of an edge line and the object relation to extract an edge line starting at the road side will be given. The used object relation that searches an edge line from a road side and a reprocessing relation with an alternative search area are shown in fig. 6 and 7.

```
DEFINE RELATION
        type:    spatially related to
        from:    road side
        to:      edge line

// part for hypotheses generation
        priority:        2
        essential:       YES

// part for goal-directed segmentation:
        image processing technique:
        - maximum cost path calcula-
        tion by dynamic programming
        parameters: 15 pixels used to
        calculate local direction of the
        road side
        search area: parallel to road
        side at distance of 3.0 m.

// part for object recognition:
// geometric conditions:
        #vec   edge   line/#vec   road
side:   [< 10]
ENDDEF
```

Fig. 6: Definition of spatial relationship (#vec = number of vectors)

```
DEFINE REPROCESSING
        type:    spatially related to
        from:    road side
        to:      edge line

// part for goal-directed segmentation:
        image processing technique:
        - maximum cost path by
        dynamic programming
        parameters: 15 pixels used to
        calculate local direction of the
        road side
        search area: parallel to road
        side at distance 1.1 m.
ENDDEF
```

Fig. 7: Definition of reprocessing relation with alternative search area for relation 6.3

Goal-directed segmentation activates a dynamic programming algorithm (Gerbrands, 1988), which detects the maximum cost path in a region of interest. The relative position of this search area is defined in the object relation using the standards for road construction (Rijkswaterstaat, 1975). The distance between the road side and the edge line differs for the left and the right side of the road (right in the sense of the side one drives at in the Netherlands). There is a hard shoulder of 3.0 m. wide at the right side of the road, while at the left side there is only a narrow correction strip of 1.1 m. wide. As a consequence, two different search areas need to be defined for searching a edge line: one parallel to the road side at a distance of 3.0, the other at a distance of 1.1 m. The reprocessing relation is added for this purpose.

The result of applying of these object relations on the aerial image in fig. 8a is shown in fig. 8b and 8c. Searching an edge line from the inner road side at a distance of 3.0 m. as specified in the object relation of fig. 6 results in the detection of the lines in fig. 8b, corresponding to object 1 and 2 in fig. 9. The objects 3 and 4 and the two lines at the outer road sides in fig. 8c also result from searching an edge line from the outer road side at a distance of 3.0 m.

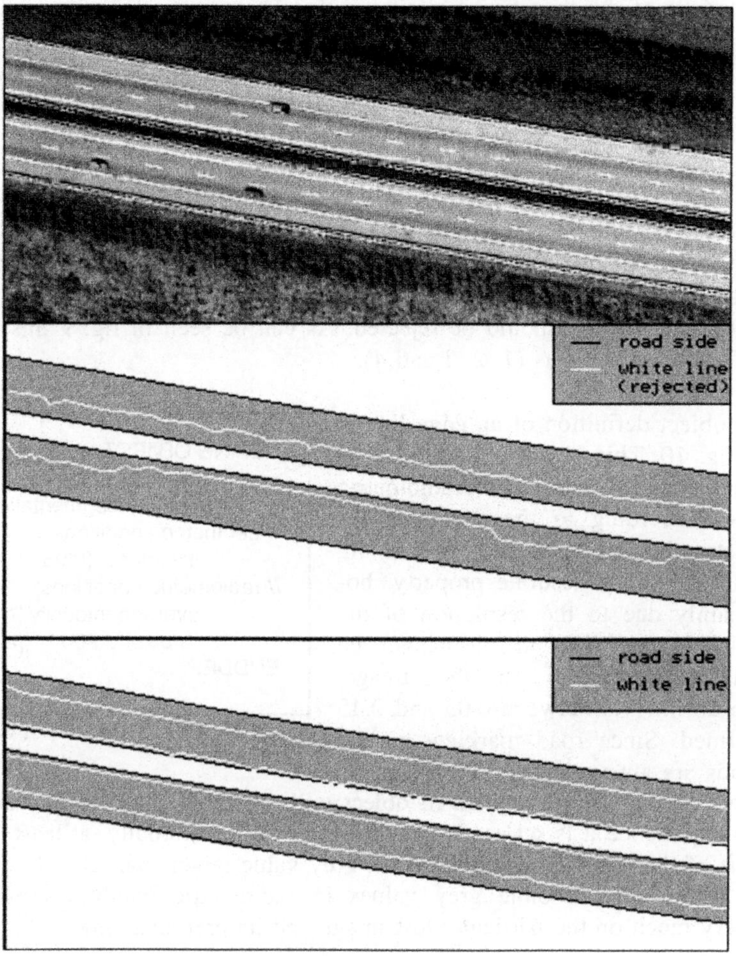

Fig. 8: (a) Aerial image with 2-lane motorway, (b) Rejected edge lines detected from inner road sides, (c) Accepted edge lines at 3.0 and 1.1 m.

	object number					
	1	2	3	4	5	6
#vec edge line / #vec road side	7	8	5	1	2	6
Average grey value	170	180	235	218	247	244
% of grey values > 200	10%	18%	90%	72%	96%	93%

Fig. 9: Values of properties of segmented objects, used for recognition of edge lines

A characteristic of dynamic programming is that always a path is found. Therefore it needs to be determined which of the two lines found at each side of the road at a distance of 3.0 m. will be accepted as edge line. For this purpose criteria are defined in the object definition of an edge line (Fig. 10). In addition one geometric property for object recognition is defined in the object relation in fig. 6, because it is related to a characteristic of the image processing technique.

Another characteristic of dynamic programming is that the path is often erratic. It defines that the detected edge line should have more than 10 times as many vectors as the road side. Since they are parallel, a much larger number of vectors indicates a very erratic path, which should be rejected. As can be seen in fig. 9 this criterion is fulfilled for all four objects (1, 2, 3 and 4).

```
DEFINE OBJECT
        name:  edge line
        origin:  segmentation
// geometric conditions:
        thickness: [0.05 m., 0.45 m.]
// radiometric conditions:
        average intensity:[≥ 200]
        % pixel ≥ 200      [≥ 70%]
ENDDEF
```

Fig. 10: Object definition of edge lines

The used object definition of an edge line is given in fig. 10. This object definition contains one geometric and two radiometric properties. According to the standards for road construction an edge line is 0.15 m. thick. This is a very accurate property, however, mainly due to the resolution of the used aerial image (0.40 m.) it cannot be measured that accurate in the image. Therefore an interval between 0.05 and 0.45 m. is defined. Since road markings in the Netherlands are always painted in white, the average grey value of the segmented object will always be high in intensity images. Based on experience it is set at a grey value of 200. Additionally, at least 70% of the pixels part of the edge line should have a grey value larger than 200. In general it is not wise to include absolute grey values in radiometric conditions because they depend very much on the particular test image and its preprocessing.

Fig. 9 shows that for object 3 and 4, corresponding to the lines at the outer road sides of fig. 8c, both radiometric conditions are fulfilled. They are accepted as edge line. However, for object 1 and 2, represented in fig. 8b, the conditions are not fulfilled. At the inner road side of a 2-lane motorway the distance between road side and edge line is actually 1.1 m. according to the standards for road construction (Rijkswaterstaat, 1975). Consequently this hypothesis was correctly rejected.

Because the object relation in fig. 6 is essential, there is a constraint that an edge line related to every road side. As a result, re-processing starts with a search area at a distance of 1.1 m. as specified in the object relation of fig. 7. It results in detection of object 5 and 6, represented at the inner road sides of fig. 8c. Fig 9 shows that all conditions for these objects are fulfilled and they are both accepted.

5.4 Result of Extraction of Road Sides and Road Markings

The result of applying the complete semantic network in fig. 5 at the aerial image of fig. 11 is shown in fig. 12. A medium scale Y-junction (in grey) is in this example extracted manually, but it could also be obtained from a medium scale digital map. This Y-junction activates the initial search for the triangular marks and road sides. The detection of these triangular marks on their turn result in search for block lines, while detection of the road sides starts the search for edge lines, which on their turn generate search actions for lane lines. The detected lane lines are rather erratic, because the maximum cost path is detected, but the distance between successive white markings of the line is rather large. This can be repaired by including an algorithm for smoothing the line and introducing knowledge about the distance between successive road markings of a lane line.

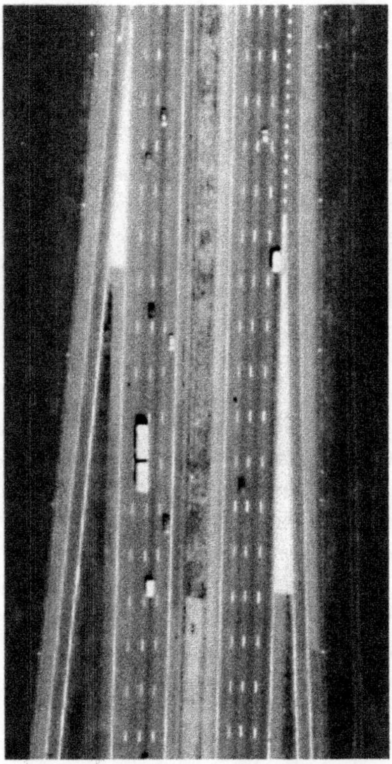

Fig. 11: Original aerial image used for extraction of road marks, road sides and white lines with contextual reasoning

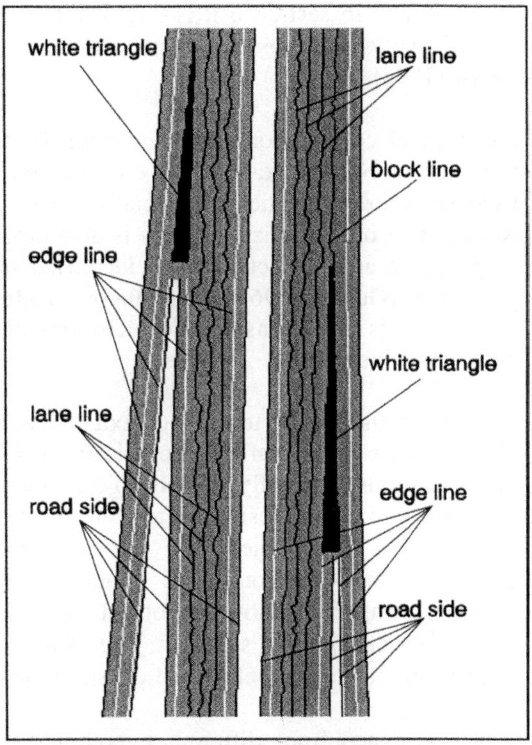

Fig. 12: Result of detecting triangular marks, road sides and white lines with contextual reasoning

6. Discussion

The contents of the knowledge base plays an important role in the performance of the interpretation strategy. Objects and their relationships are the basic primitives, represented in the knowledge base by frames. Object definitions are used for the task of object recognition in the interpretation strategy. Object relationships are used for hypotheses generation, goal-directed segmentation and inconsistency detection. The question is, which parts of the road network should be defined as objects and which properties should be used to build a road model. The complex contents of aerial images requires recognition of meaningful parts of the road network instead of linear structures. Therefore, the knowledge base used in this paper is part of a framework for a complete road network model, covering multiple scales of imagery. The connection between various scales reflexes the complex composition of parts of road networks. In order to handle a certain application, the knowledge base at a certain level of detail can be created by definition of specialized object classes and spatial relationships. In section 6 it is shown that recognition of meaningful objects at large scale is very well possible. Applications at small and medium scale are given in Gunst (1996).

An original contribution is the insertion of standards in the road model, which are used in the Netherlands for the construction of road networks. These standards are used to define properties of specialized road types and relationships between roads. Recognition of specialized classes is also profitable if only general classes need to be mapped, because objects can be identified more reliably with these specific object properties (Gunst, 1996). When these standards are incorporated in the knowledge base, it needs to be considered which properties to use for search, recognition and inconsistency detection.

Inherent to the choice to use contextual relationships to search new objects is that a chain of successive search actions is formed. As a consequence omission errors also cause that other spatially related objects are not found. Therefore, one should watch carefully during the design of an interpretation strategy that there is more than one way to find every object and that re-segmentation is done. There is no general procedure for adaptation of an image processing technique, its parameters, search area or using an alternative object definition. Which alternative hypotheses are defined depends on the situation. The example in section 6 shows clearly that knowledge from the standards for road construction can be used for this purpose.

When evidence from different hypotheses is obtained, the current mechanism for handling inconsistencies by means of essential relationships is too simple. It will be necessary to include reasoning with uncertainty. Vosselman (1996) gives an overview of different ways of representing uncertainty in data and processing of uncertain data. However, the problem is that probabilities are needed to describe the uncertainty in the model, but these probabilities and the way in which they propagate during processing are both unknown.

The knowledge based in is paper is restricted to the semantic road model. However, also other topographic objects and non-topographic objects like cars yield information relevant for the extraction of roads. As already stated in the introduction of this paper, building a knowledge base which includes all topographic objects and their relationships is a very difficult task. Besides, the interpretation strategy resulting from a knowledge base with many objects will be difficult to control. Therefore, we advice to use the concepts for limited interpretation tasks.

Acknowledgements

The research project was financially supported by TNO Institute of Applied Physics, Delft, and the Survey Department of Rijkswaterstaat, Delft.

References

Bajcsy, R., Tavakoli, M. (1976) *Computer Recognition of Roads from Satellite Pictures*, IEEE Transactions on Systems, Man and Cybernetics, Vol. SMC-6, No. 9., pp. 623-637.

Cleynenbreugel, J. van, Fierens, F., Suetens, P., Oosterlink, A. (1990) *Delineating Road Structures on Satellite Imagery by a GIS-Guided Technique*, Photogrammetric Engineering and Remote Sensing, Vol. 56, No. 6, pp. 893-898.

Draper, B.A., Collins, R.T., Brolio, J., Hanson, A.R., Riseman, E.M. (1989) *The Schema System*, International Journal of Computer Vision, Vol.2, pp.251-282.

Fischler, M.A., Tenenbaum, J.M. Wolf, H.C. (1981) *Detection of Roads and Linear Structures in Low-Resolution Aerial Imagery Using a Multisource Knowledge Integration Technique*, Computer Graphics and Image Processing, No. 15, pp. 201-223.

Gerbrands, J.J. (1988) *Segmentation of Noisy Images*, Ph.D.thesis, Delft University of Technology, The Netherlands, 170 p.

Groch, W. (1982) *Extraction of Line Shaped Objects from Aerial Images Using a Special Operator to Analyze the Profiles of Functions*, Computer graphics and image processing, pp. 347-358.

Gunst, M.E. de (1996) *Knowledge-Based Interpretation of Aerial Images for Updating Road Maps*. PhD-thesis, Delft University of Technology, Faculty of Geodetic Engineering, 184 p.

Hartog, J.E. den (1995) *A Framework for Knowledge-Based Map Interpretation*. PhD-thesis, Delft University of Technology, Faculty of Electrical Engineering, 88 p.

Matsuyama, T. (1987) *Knowledge-Based Aerial Image Understanding System and Expert Systems for Image Processing*. IEEE Transactions on Geoscience and Remote Sensing, Vol. GE-25, No.3, May 1987, pp. 305-316.

Matsuyama, T., Hwang, V. (1985) *SIGMA: A Framework for Image Understanding - Integration of Bottom-Up and Top-Down Analysis*, International Joint Conference on Artificial Intelligence, pp.908-915.

Rijkswaterstaat (1975) *Standards for the Construction of Motorways* (in dutch: Richtlijnen voor het ontwerp van autosnelwegen), Ministerie voor Verkeer en Waterstaat, Dienst Verkeerskunde (DVK), current name: Adviesdienst Verkeer en Vervoer (AVV)

Vosselman, G. (1996) *Uncertainty in GIS Supported Road Extraction.* International Archives of Photogrammetry and Remote Sensing, Proc. of XVIII ISPRS Congress, Comm. III, Vienna, Vol.31, Part B3, pp. 909-917.

Semanitic Modeling for the Acquisition of Topografic
Information from Images and Maps
ed. by W. Förstner and L. Plümer

Road Modeling Based on a Cartographic Database for Aerial Image Interpretation

Ghislaine BORDES *, Gérard GIRAUDON **, Olivier JAMET *

* IGN / MATIS
2, avenue Pasteur
94165 - Saint-Mandé Cedex - France
Ph : +33 1 43 98 80 00, Fax : +33 1 43 98 85 81
email : Ghislaine.Bordes@ign.fr

** INRIA - France

Abstract

The need of semantic models for topographic objects detection from images has been demonstrated in many works. Existing databases or maps may provide semantic knowledge on topographic objects of the scene. Thus they facilitate the semantic modeling. In this paper, the problem of road modeling is considered. An existing cartographic database is used to build a semantic image-independant road model : the DB semantic road model. An adaptive road model for the road detection is deduced from the DB semantic model. That is that the DB semantic road model is used to select the most appropriate road model among several ones for the detection of each road section, depending on the road characteristics and context. A road extraction system based on the DB semantic road model and on different detection road models has been designed and tested. Some results of road extraction from aerial images are presented. The efficiency of this adaptive road model is discussed in the last part of the paper.

1. Introduction

Capturing topographic information automatically from aerial imagery is a very complex task. The topographic objects to be detected (roads, buildings...) have very various aspects in the image, depending on their own characteristics and on their context. In other words the main difficulty is to define models for the topographic objects. Detection operators detect objects which fit a precise model. If the topographic object as it appears in the image does not correspond to the model, the operator fails in detecting it. The problem is to define models which are generic enough to reach a good rate of completeness but precise enough to avoid false detections. In manual detection, the "modeling" is based on the operator experience and involves high level semantic knowledge. It appears that automatic processes also require fine semantic knowledge about topographic objects. One of the simplest solutions to build semantic models is to use external knowledge provided by an existing map or by a database. This solution is studied in our paper.

The objective of our research is the automation of road extraction from aerial images using a Cartographic Database (CDB). The long term objective is the acquisition of a topographic database which accuracy is 1m. The aerial images are ortho-images with a ground resolution of 50 cm per pixel. Our problem is not to update the Cartographic Database, which is very inaccurate (about 20 m in planimetric standard deviation) compared to the image resolution, but to use it as a guide for the interpretation. The Cartographic Database provides semantic information on the scene and on its topographic objects. Thus, it is a good tool to refine the objects models or to choose the best model among several ones.

Section 2 of this paper gives a brief overview of road models used in classical road extraction. Then, the semantic road model derived from the Cartographic Database is described. In section 3 our application is presented : the principles and the architecture of our road extraction system are described. In section 4, some results are presented and discussed.

2. Road modeling

2.1 Generic road models

Any road extraction algorithm is based, explicitly or not, on a road model. Generic road models used in previous works are presented by De Gunst (1991 and 1996). These models are linked to the characteristics of the imagery (resolution in particular) and to the characteristics of roads which have to be detected. Most often the road model chosen would not be suitable for an other type of road or in an other type of landscape. On the contrary, if the road model is too generic other objects such as rivers or railways may be detected.

The main reason of this difficulty is that road models have to express the fuzzy knowledge which is usually used by an operator for road extraction. For instance, models representing roads as parallel edges cannot use an exact definition of parallelism but must have a tolerance value on the difference of edges directions. Thus, such fuzzy properties are expressed by fuzzy functions or by tolerance thresholds which are difficult to define. On the other hand, many properties such as size of roads, their radiometry...can be valuated. But this valuation cannot either be exact, since these values may vary a lot from a type of road to an other, from a type of landscape to an other. Most classical road models only provide average values or extremum values.

To avoid these problems of valuation, some road models use only relative properties of roads : "the road width is constant", "the road profile varies very gradually" (MacKeown, 1988), "the radiometric variance is low along the track of the road" (Airault, 1994)...These "relative" road models are used by road following algorithms that require road seeds provided by an operator or by an automatic detection. The road model adapts itself to the seed point characteristics and is modified gradually during the following.

Through these previous works it has appeared that no simple generic model can be defined for automatic road extraction. Recent works raise more directly the problem of modeling and define more complex road models which take into account high level semantic knowledge. Garnesson (1991) define "semantic objects" such as roads according to 4 points of view : the radiometry, the geometry, the context (interactions with other objects) and the function. This 4 levels model is intended for various applications, various images...It is a high level semantic model which uses fuzzy sets and possibility theories. For instance, objects size is described by a possibility distribution. As far as radiometry is concerned, objects are described by their material which are related to fuzzy adjectives (bright, dark...) corresponding to their aspect in the image. Spatial context of roads describes their possible links with other semantic objects. These links have an uncertainty factor which qualifies their probability to be verified. This solution is based on a high level semantic description of the scene completely independant of the image. It involves models for the objects which have to be detected but also for other related objects, for the image itself and for its acquisition process.

Our proposal to simplify the problem of road modeling is to use external data provided by a cartographic database.

2.2 Road model derived from the Cartographic Database

In this work, a Cartographic Database (CDB) is available in addition to the aerial image in order to perform the road extraction. This Database corresponding roughly to 1:100 000 maps is produced by the French I.G.N[1]. It is already available for the whole territory and is updated frequently. It contains the main communication and hydrographic networks and landcover areas. All these cartographic objects are represented by vector data and have semantic attributes. Remember that our objective is not to update the Cartographic Database road network - which is updated frequently in any case - but to find in the image the precise location of the roads which are in the database. Therefore we do not take into account roads of the image which are not in the database and consider that a road which is in the database is always present in the image.

We define a road model which gathers all the database information useful for the road detection in the image. This road model derived from the CDB will be called the "DB

1 National Institute of Geography

road model". Roads are represented by polylines called "DB Road Sections". Road sections are connected at "DB Crossroads" (represented by a point) which correspond to real crossroads of the road network but may also correspond to a simple change of road characteristics. A DB Road Section has semantic attributes which describe the inner characteristics of the road and its context.

DB Road Section	**DB Crossroads**
- geometric description (polyline)	- geometric description (point)
- initial crossroads	- type : { simple, traffic circle, complex intersection }
- final crossroads	- number of sections connected
- number of lanes	- sections connected
- number of separated carriageways	
- type of road {freeway, main road, secondary road, lane, footpath}	
- position on the ground	
- land cover : { urban, forest, others}	
- neighboring linear objects : {railway, river}	

Fig. 1 : DB Road model

The instances of these different objects are created from the CDB. They can be projected onto the aerial ortho-image (see Fig. 2). They describe the scene : they provide the rough location of roads in the scene and semantic information for each of these roads.

The DB model is an image-independant semantic road model. Each instance of DB Road Section provides a *specific model* for each road which makes it possible to predict the road aspect in the image :

- The "geometric description" of the DB road section gives information about the road shape (curvature in particular) and length.
- Semantic attributes "number of lanes", "number of separated carriegeways" and "type of road" gives information about the road width, about the presence of road marking and about the number of parallel edges that can be detected in the image.
- Attributes "land cover" and "neighboring linear objects" provide contextual information about objects that can be found in the neighborhood of the road.

The DB model is a limited model in the sense that some cartographic objects which are related to roads such as buildings or trees are not represented as such. On the other hand, the DB road instances are inaccurate compared to the image resolution. All these drawbacks have to be taken into account in the interpretation strategy.

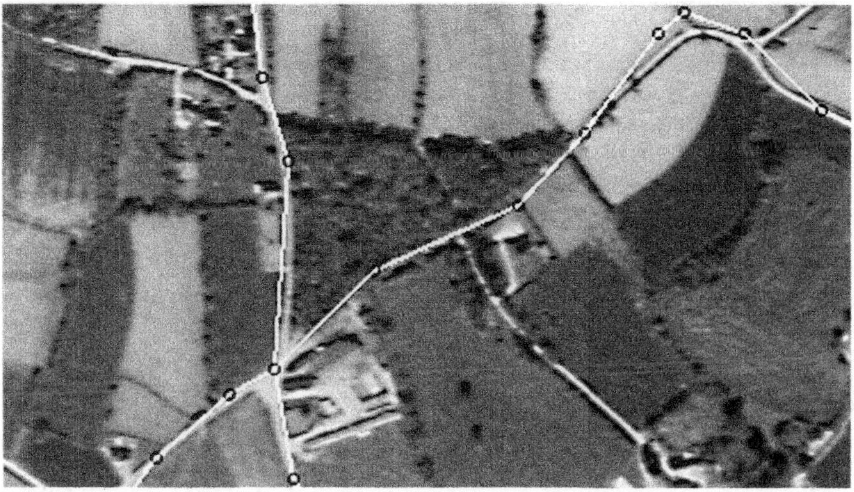

Fig. 2 : DB roads (in white lines) projected onto the ortho-image

2.3 Road extraction models

The DB Road model represents declarative knowledge which is not directly usable for the road extraction in the image. To process the road extraction, the DB road model has to be "translated" into image road properties. Two types of properties are used in road extraction models (see Fig. 3) : generic properties (radiometric, geometric, topologic) of roads in images and properties related to the DB Road sections. These properties are grouped into different *road extraction models* which correspond to different types of roads. A road detection model can be expressed by a set of properties : {GP1, GP2....DB1, DB2...}

In practice, the road extraction models are not represented by structured objects (as opposed to the DB model) but through procedures. These procedures are road extraction algorithms based on the model road properties.

Thus, different classes of DB road sections are defined according to their attributes and each class is associated to the most consistent road extraction model (see Fig. 6). On the other hand, possible parameters of the model, such as width of road, are tuned according to the DB road attributes. Then, the road extraction algorithm associated to the road model can be computed and provides road primitives.

Generic properties :

- **GP-Color :** roads are light-coloured linear features on a dark background.
- **GP-Variance :** roads have a low radiometric variance along their track. The variance possibility distribution is described by the fuzzy function $F_{Variance}$ (see Fig. 4).
- **GP-Edges :** roads have contrasted parallel edges.
- **GP-Profile :** the radiometric profile of roads varies very gradually.
- **GP-Width :** roads width is constant and the road width possibility distribution is defined by the fuzzy function F_{Width}.
- **GP-Length :** roads have a minimal length.
- **GP-Shape :** roads have a low curvature, except at crossroads. The"linearity" possibility distribution is described by the fuzzy function $F_{linearity}$ (see Fig. 4).
- **GP-Network :** roads form a connected network.

Properties related to the DB :

- **DB-Direction :** road average direction is close to the DB Road section direction. The angular divergence possibility distribution is defined by the fuzzy function $F_{Divergence}$ (see Fig. 5).
- **DB-Shape :** road shape is close to the DB road section shape. The length of road is close to the DB Road section length. The possibility distribution for the values of the ratio (road length)/(DB section length) is defined by function F_{Length}.
- **DB-Proximity :** roads are situated in the neighborhood of DB Road sections.
- **DB-Width :** road width depends on the DB road characteristics and possibility distributions for road width are defined for each type of roads (same type of fuzzy functions as F_{width} but with more specific width values).

Fig. 3 : Road properties

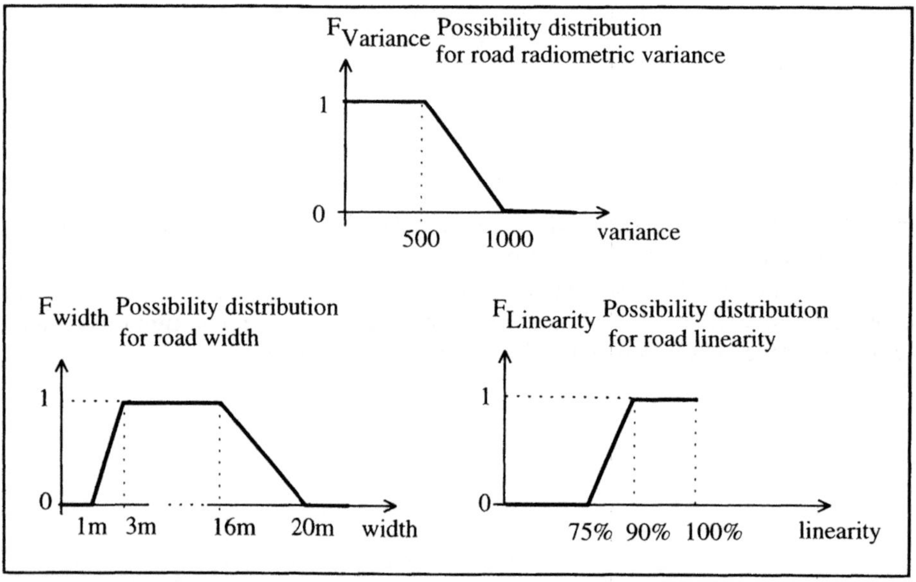

Fig. 4 : Possibility distributions for road generic properties

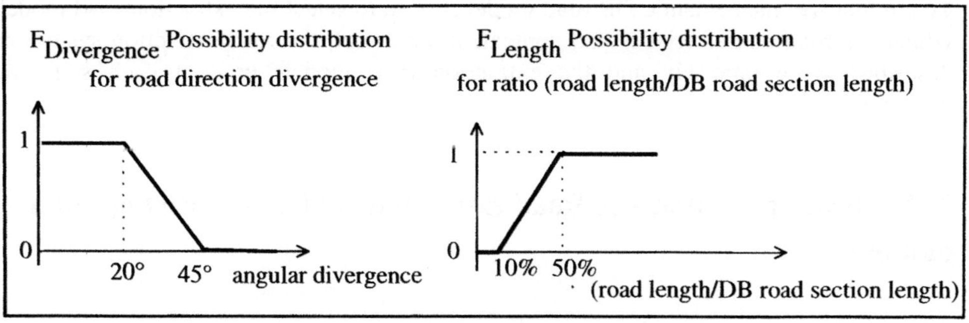

Fig. 5 : Road extraction models and the DB road model

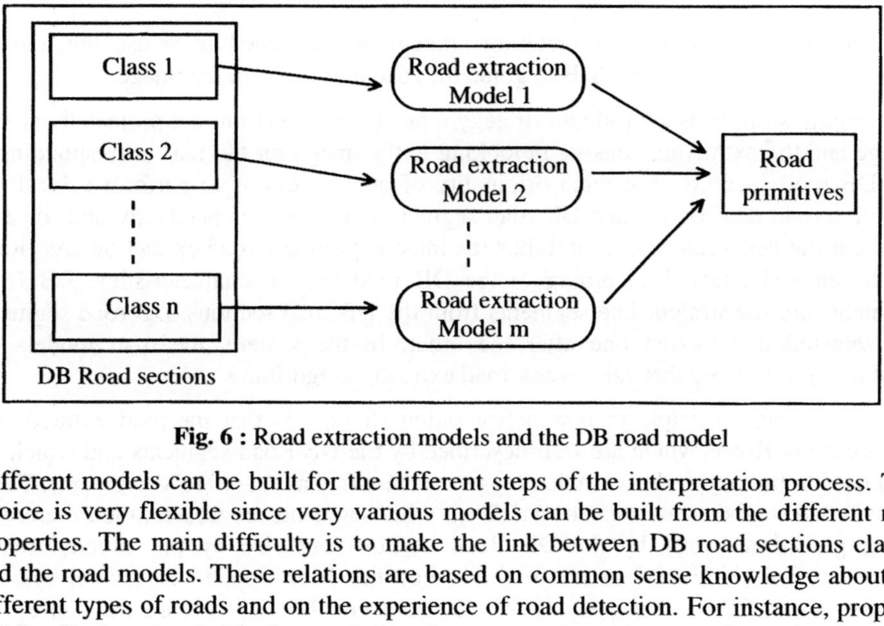

Fig. 6 : Road extraction models and the DB road model

Different models can be built for the different steps of the interpretation process. This choice is very flexible since very various models can be built from the different road properties. The main difficulty is to make the link between DB road sections classes and the road models. These relations are based on common sense knowledge about the different types of roads and on the experience of road detection. For instance, property GP-Profile is not suitable for roads in urban areas, since these roads are bordered by small objects such as buildings or cars which affect their profiles.

Thus, our solution avoids the problem of defining a generic road model by using several models and selecting the most appropriate one for each road. The same type of solution was proposed by MacKeown (1988) : two cooperative road trackers based on complementary models are used. But in this previous work, both trackers were processed for each road and the choice between the two road tracks was made a posteriori. In our case, the DB road model makes it possible to choose the best model a priori using high level semantic knowledge. Thus various road models can be used without increasing the computing time.

To summarize, the semantic DB road model is used to select the extraction road model which fits road characteristics and context. In next part, the road extraction system is described more precisely and the extraction road models used at each step are presented.

3. Application : Automatic Road extraction guided by a cartographic database

3.1 Interpretation strategy

The general idea of the interpretation strategy we designed is to use the a priori knowledge provided by the CDB to guide the road extraction in the image.

The overall strategy is a top-down strategy : the DB road sections are projected onto the image and the extraction consists in looking in the image for the road corresponding to the DB road sections. The main originality of our strategy is to perform a local and adaptive road detection : each DB road segment is treated independantly and for each segment the best road model (and thus the most appropriate road extraction algorithm) is chosen and adapted according to the DB road section characteristics. DB Road segments are the straight line segments from the DB road sections. DB road segments are selected and treated one after the other. In the system, the road models are implicitely expressed through several road extraction algorithms.

The other main principle of our interpretation strategy is that the road extraction is hierarchized. Roads which are well-described by the DB Road segments and which are supposed to be easy to detect in the image are classified as "easy" road segments. These "easy" road segments are treated first. Then other segments are treated, their extraction being guided not only by the DB Road sections but also by the already treated segments. This kind of "easiest first" strategy has been applied in several systems (Garnesson, 1991 and Sandakly, 1995, for instance). But in our case the concept of easiness is related to the DB Road section characteristics and location and not only to the road aspect in the image.

As a result, this interpretation strategy requires two types of control knowledge :

- Rules to select the most appropriate road extraction algorithm (associated to a model) for each DB road segment and to tune its parameters. These rules are based on common sense knowledge and on experience about extraction algorithms performances.
- Rules to predict the easiness level of each DB road segment. This easiness level takes into account the a priori reliability of the DB road sections and image tokens calculated for end points and midpoint of the DB segment. The a priori reliability is deduced from DB section attributes, it reflects the reliability of the DB Road

section location, knowing that it has been generalized and simplified during the CDB acquisition. The prediction of reliability (see Bordes, 1995) is based on the cartographic generalization rules which take into account the significance of the road. As far as image tokens are concerned they provide information about the road aspect in the image. They are simple measures based on very simple road models. Their computation consists in verifying if the DB Segment endpoints and midpoints projected onto the image fit to the different road models used by the road extraction algorithms. In our test-system - which is a simple and limited implementation of the strategy presented above - only two image tokens are used currently The first one is based on radiometric homogeneity (Property GP-Variance) and the second one is based on the radiometric profile characteristics (Property GP-Profile). Both tokens suppose that the road direction is close to the DB segment one (Property DB-Direction) and that the road is located in the neighborhood of the DB segment (Property DB-Property). So called "easy road segments" are reliable and have good image tokens.

3.2 System architecture

The interpretation system is based on a blackboard architecture (see Fig. 7). In this type of systems, declarative knowledge is stored in the blackboard. Procedural knowledge is described by "knowledge sources" which can read or write information in the blackboard. This kind of organization is interesting for our strategy since it separates clearly knowledge and processings. Thus it is possible to add easily new road extraction algorithms corresponding to different road models. On the other hand, all the data stored in the blackboard can be read by any knowledge sources at any time and this condition is necessary for our interpretation strategy.

The blackboard contains semantic objects such as DB Road sections and low level objects such as road primitives and images. The objects are structured and linked with others. Images are described by their resolution, their cartographic coordinates and their grey level pixels. Primitives are polylines which correspond to roads detected in the image.

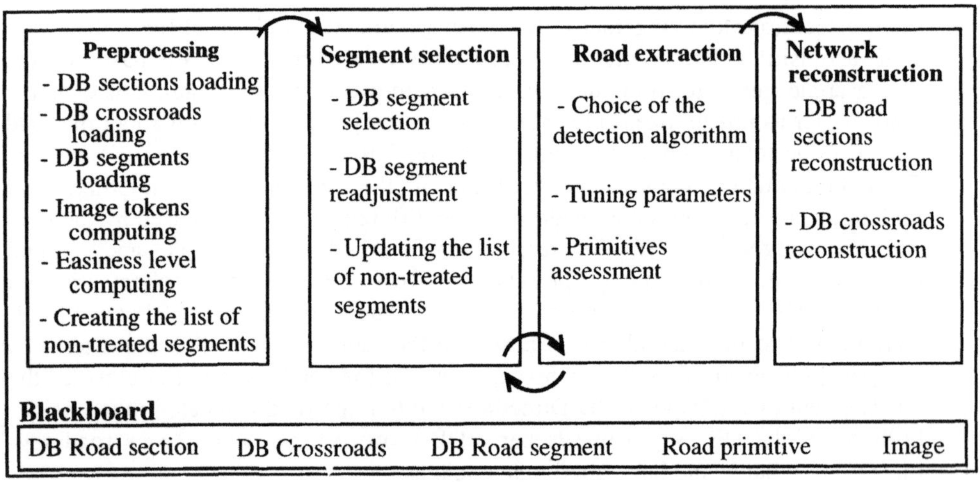

Fig. 7 : System architecture

Preprocessing : The first step of the processing consists in loading DB Road sections and DB Crossroads from the Cartographic Database. DB road segments are deduced from DB road sections, their image tokens and their easiness level are calculated. This preprocessing provides the list of DB road segments to detect in the image.

After this preprocessing step, the detection is an iterative process : a segment is selected in the list of non-treated DB road segments, the road extraction is performed for this segment and the DB road segment is eliminated from the list of non detected segments, then the next segment is selected etc....

Segment selection : At each step, the segment to detect is selected according to several criteria (by priority order) : the road context, the segment easiness level, the connectivity to already treated segments and the length of the segment. More precisely, road segments which are situated in simple context and which are classified as easy segments are selected first, then if the remaining segments have the same easiness level, the priority is given to segments connected to already-treated ones. This selection order corresponds to a "propagation" which relies on easiest road segments to detect less easy ones. When the segment has been selected, its endpoints and midpoint are adjusted on the road axis. Several adjustment methods can be used, depending on the segment characteristics. If the segment is an easy one, a simple adjustment by searching parallel edges is performed. In other cases, the segment is firstly adjusted on road primitives detected at low resolution by a top-hat operator and eventually adjusted more precisely. If the segment is connected to an already extracted primitive, the connection point is simply adjusted on the primitive endpoint. The quality of the adjusted points of the segment is stored. Points are considered as "well-adjusted" only if they have been adjusted on well-contrasted parallel edges or by connection to an already extracted primitive.

Road extraction : Then, the road extraction is performed - for the selected DB road segment. Firstly, a road extraction algorithm is chosen. Three algorithms are currently available : a road following by profile analysis, a road following by homogeneity criteria and a simple connection of the well-adjusted points of the DB road segment. The third algorithm is chosen only for very short segments, the points of which are well-adjusted. This algorithm is based essentially on properties related to the DB, the underlying model is {GP-Shape, DB-Direction, DB-Shape, DB-Proximity}.

In other cases, only the road followings can be used. The road following by homogeneity criteria is based on the model {GP-Variance, GP-Edges and GP-Shape}. This algorithm is described by Airault (1994). The road following by profile analysis is based on profile matching and uses the model {GP-Profile, GP-Shape}. This type of algorithm is described by MacKeown (1988). These two road following algorithms require a road seed and a propagation direction. The road seed is one of the well-adjusted points of the DB segment. The best adjusted point is selected, according to its image tokens. If none of these points is correctly adjusted, no seed is selected, the extraction stops and the next segment is selected. Otherwise, when the road seed has been selected the most appropriate road following is selected depending on the image tokens of the seed or on the road section attributes. If the seed point is connected to an already extracted primitive, the system selects the algorithm which has been used for the extraction of this primitive.

The propagation direction is the DB segment direction (Property DB-Direction). Other parameters(stop criteria...) are tuned according to the image tokens and to the segment characteristics. If the segment is connected to an already extracted primitive, the ingoing and outgoing parameters used for the extraction of this primitive are used for the detection of the current segment.

The road followings are processed only in the neighborhood of the DB road segment. Resulting primitives are stored with their features (see Fig. 8). On the other hand, for each DB Segment, the detection parameters are stored in "memory tables" (see below "Detection memory" and "Road-following-memory"). These tables are used to control the system : firstly for segments connected to an already extracted primitive, they allow to use the outgoing parameters stored in the memory tables, secondly they can be used to analyze the possible failures in the detection and to perform an alternative detection.

Then, the primitives generated for the current DB segment are assessed by comparison with two models. The assessment method is based on fuzzy sets and possibility theory, the same type of method has been used in the MESSIE system (see Sandakly, 1995). The first assessment model takes into account generic properties {GP-Variance, GP-Shape, GP-Width}. The generic quality of primitives is calculated by comparing the different features (variance, linearity and width) of the primitive to the corresponding possibility distributions defined in the image model. The second quality assessment is based on DB properties of the model {DB Direction, DB-Shape}. The primitives features average direction and length are compared to the model possibility distributions $F_{Divergence}$ and F_{Length} and combined.

Road primitive	Detection-memory
- identification - geometric description (polyline) - related DB Segment - detection algorithm { connection, road following by homogeneity, road following by profile analysis, top hat operator} - detection resolution - length - linearity - mean width - radiometric mean - radiometric variance - generic quality [-1; 1] - specific quality [-1; 1] - quality level {0, 1, 2, 3, 4}	- DB segment - DB section associated - road seed {no seed point, initial pt, final pt, midpoint} - detection algorithm - result {0, 1,...primitives} - readjustment by top hat {yes, no} - road following ingoing parameters (only for failures) **Road-Following-memory** - primitive identification - ingoing parameters - outgoing parameters

Fig. 8 : Primitive structure and memory tables

Network reconstruction : When the list of non-treated segments is empty, the road network is reconstructed. Firstly, DB road sections are reconstructed. All the primitives depending on the segments related to the same DB section which have a good quality level are connected. For the moment, this connection is a simple geometric connection of the primitives endpoints. But it could be improved in future work and take the image into account. These connection parts are stored as primitives which are called "interpolated primitives".

When all DB road sections are reconstructed DB crossroads are reconstructed by connecting the incoming and outcoming primitives. Only simple crossroads are correctly reconstructed for the moment. More complex crossroads have not been studied.

4. Results

This automatic road extraction system has been tested on several images. Results in terms of accuracy and of completeness have been assessed. The accuracy is calculated by automatic matching of the road primitives points with a reference road network manually plotted and the accuracy of which is about 1m. The completeness is evaluated by comparing the road primitives length to the DB road length. Each primitive point which is matched provides a deviation measurement. Statistical values such as Root Mean Square are computed from these measurements.

The results of road detection are presented in Fig. 9 for a 13 km^2 image. The image resolution is 0.45m and the total length of the DB road network for the same area is about 20 km. Validated primitives are the ones which have been detected at high resolution (and not during the network reconstruction) and which quality level is good (greater than 2).

	Number of measurements	RMS_{XY}	Completeness
DB road sections	2130	7.2m	100%
Validated road primitives	1054	1.6m	67%
Interpolated primitives	389	5.6m	33%

Fig. 9 : Detection results

It appears that the accuracy of validated primitives is good compared to the initial accuracy of DB road sections. Nevertheless, the Root Mean Square required for the result ranges from 1 to 2 meters, the RMS of validated primitives is a bit too high. This poor accuracy is essentially due to false detections in urban areas. The analysis of results according to the land cover area attribute shows that in urban area the RMS is 1.4m whereas in rural areas it is 2.1m. In urban areas, DB road sections and crossroads are strongly simplified (see Fig. 14, 15). Thus it is more difficult to find "easy" road segments to initiate the detection. On the other hand, the adjustment process for segment endpoints is not reliable in urban areas, and may provide false road seeds. For rural landscapes (see Fig. 10, 11, 12, 13), results are satisfying.

The completeness of results (considering validated primitives only) depends on the type of landscapes and images. For rural landscapes, the completeness can reach more than 90%. But for more complex areas, such as urban areas, it can be lower than 50%. One of the advantages of our system is that this completeness level can be assessed automatically by the system. In an operationnal application, areas with a low completeness could be identified and interpreted manually.

The accuracy of interpolated primitives is not satisfying. This result is due to the simple connection method. The reconstruction stage should be improved in future work, in particular by using more complex models for the crossroads reconstruction.

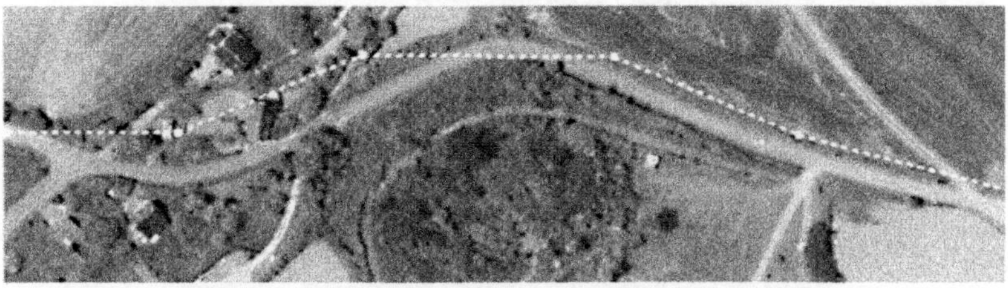

Fig. 10 : Image + DB Road section (white dotted line)

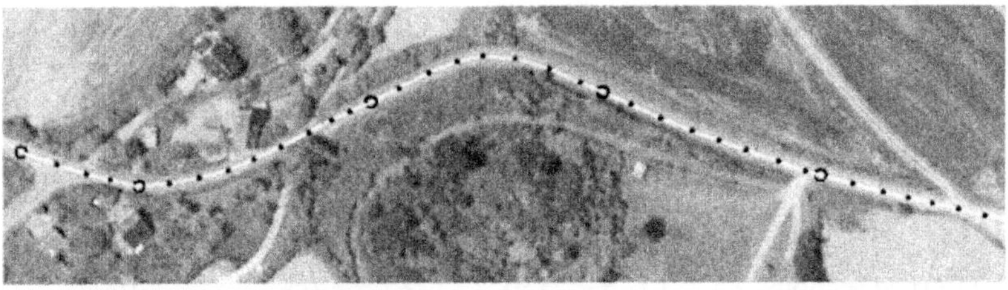

Fig. 11 : Image + Road primitives (in white) extracted for the road section presented in Fig. 10. The road extraction has been propagated from the rigth DB road segment which is classified as the easiest one for this road section.

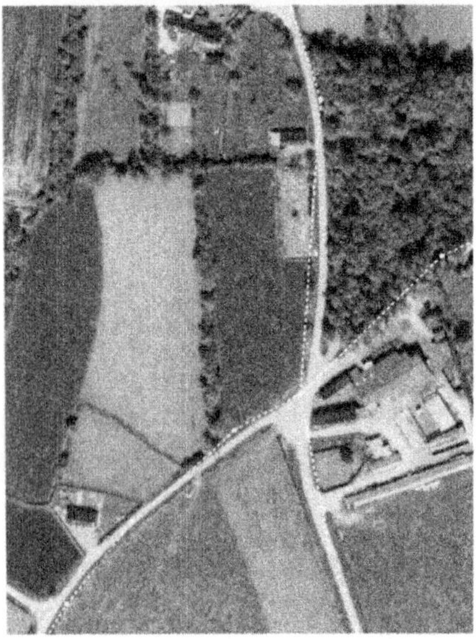

Fig. 12 : Image + DB Road network (white dotted lines)

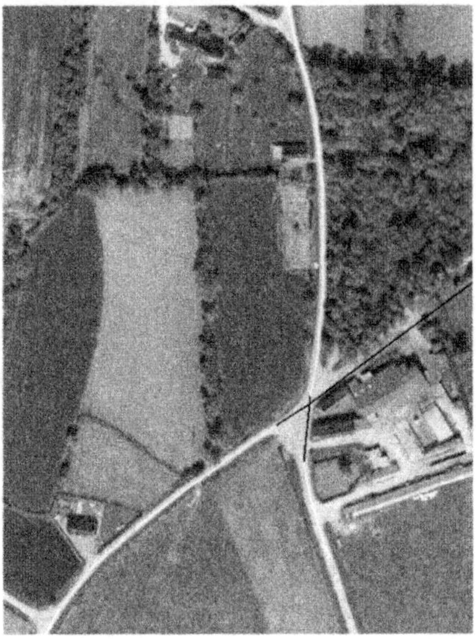

Fig. 13 : Image + Road primitives extracted and validated (in white) or interpolated (in black)

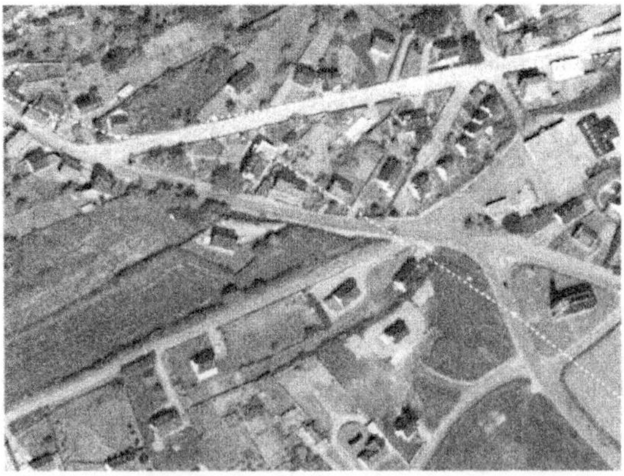

Fig. 14 : Image in urban area + DB Road section (white dotted line)

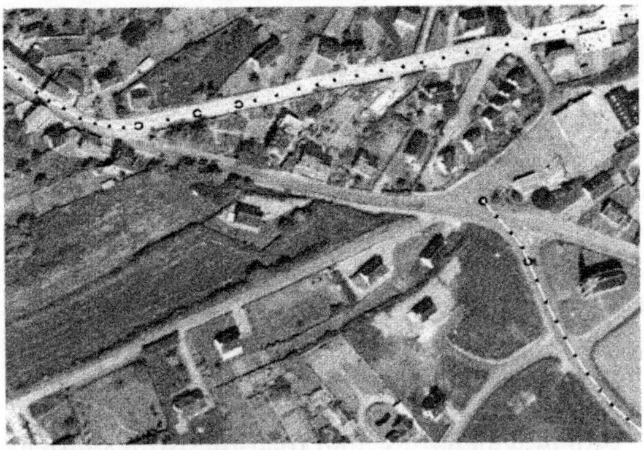

Fig. 15 : Image + Road primitives (in white) extracted for the road section presented in Fig. 14.

We have also computed tests to verify the advantages and efficiency of our models. Firstly, we studied the efficiency of the quality assessment model by analyzing the accuracy values depending on the primitive quality level (cf Fig. 16). Only extracted road primitives have been taken into account.

Primitives Quality level	**Number of measurements**	**RMS_{XY}**
Level 4	816	1.5m
Level 3	238	1.9m
Level 2	41	2.3m
Level 1	4	5.5m

Fig. 16 : Road primitives accuracy depending on their quality level

It appears that the quality level is relevant. Primitives with a high quality level are better located than the others. This result shows that the quality assessment model is efficient and could be used to distinguish automatically false detections. This result is essential on an operationnal point of view.

Secondly, we tried to prove that the use of different models is efficient in terms of completeness and accuracy of results. Thus we compared results provided by the system to the results provided by the same strategy but with a single road following.

The different results (for validated primitives only) are presented in Fig. 17. The first line presents the results of the system using both road followings. The second line gives the results when the system uses only road following by profile analysis. The third line gives the results when using only the road following by homogeneity criteria.

	Number of measurements	RMS_{XY}	Completeness (%)
System	1054	1.6m	67%
Road following by profile analysis	855	1.4m	55%
Road following by homogeneity	980	1.7m	60%

Fig. 17 : Results with a single road following

It appears that our system which uses both road followings provides the best results in terms of completeness and makes a good compromise between completeness and accuracy. This result highlights the advantages of using several road models (here represented by different road followings). It also proves that the control strategy used to select the appropriate algorithm for each road segment is relevant. Our method could be enriched by using other road models and road extraction algorithms.

5. Conclusion

In this paper, we presented a road modeling which uses knowledge from a cartographic database. From this database is derived a semantic road model which stores high semantic and image-independant knowledge. This semantic model is instantiated thanks to the cartographic database. The semantic road model instances are used to select the most appropriate extraction model for each road. This selection is performed according to the road characteristics and geographic context described by the attributes of the semantic road model. The road extraction system which is based on this "adaptive model" has been described. Several road extraction algorithms based on different road models are available.

The results are quite satisfying in terms of completeness and accuracy. The reliability of road primitives is automatically assessed by the system. This assessment has been proven to be relevant. The advantages of using these adaptive models instead of a single road model have been highlighted. The main originality of our method, compared to previously published works, is the choice of the most appropriate road model (and extraction algorithm) which is made a priori using the semantic road model derived from the cartographic database.

In future work, other road models could be added to the system, in order to improve the results for complex areas (urban areas for instance). The network reconstruction should also be improved.

This road extraction system is part of a project which has strong operationnal constraints on the result accuracy and reliability and on the ingoing images and data.

But the road modeling and the interpretation strategy which is proposed could be adapted for other applications in which external data such as maps or databases are available.

References

Airault S. (1994) Détection et restitution automatiques du réseau routier sur images aériennes. RFIA, Vol 1, pp519-531, Paris.

Baumgartner A. et al. (1996) *Update of roads in G.I.S. from Aerial Imagery : Verification and Multi-Resolution Extraction.* International Archives of Photogrammetry and Remote Sensing. vol XXXI, Part B3, Vienna.

Bordes G., Giraudon G., Jamet O. (1995) *Road Extraction Guided by a Cartographic Database : Creating a Strategy.* ICCV 95 - Workshop on Context Based Vision, Cambridge (MA).

Garnesson P. (1991) *MESSIE : Un système d'analyse de scène, application à l'imagerie aérienne.* PhD thesis, Nice-Sophia Antipolis University.

De Gunst M. (1991) *Automatic Road Extraction of Roads from SPOT Images.* Delft University of Technology, Faculty of Geodesy. Delft, Netherlands.

De Gunst M. (1996) *Knowledge-based interpretation of aerial images for updating of road maps.* Delft University of Technology, Faculty of Geodesy. Delft, Netherlands.

MacKeown D., Denlinger J. (1988) *"Cooperative Methods for Road Tracking in Aerial Imagery"*, IEEE proceedings on computer vision and pattern recognition, Ann Arbor, pp 662-672.

Sandakly F. (1995) *Contribution à la mise en oeuvre d'une architecture à base de connaissances pour l'interprétation de scènes 2D et 3D.* PhD thesis, Nice-Sophia Antipolis University.

Steger C. et al. (1995) *Model-Based Road Extraction form Images*, Automatic Extraction of Man-Made Objects from Aerial and Space Images. Monte Verità, Birkhaüser Verlag Basel.

Semanitic Modeling for the Acquisition of Topografic
Information from Images and Maps
ed. by W. Förstner and L. Plümer

Multi-Resolution, Semantic Objects, and Context for Road Extraction

A. Baumgartner[1], C. Steger[2], H. Mayer[1], W. Eckstein[2]

[1]Lehrstuhl für Photogrammetrie und Fernerkundung
Technische Universität München
E-mail: {albert|helmut}@photo.verm.tu-muenchen.de

[2]Forschungsgruppe Bildverstehen (FG BV), Informatik IX
Technische Universität München
E-mail: {stegerc|eckstein}@informatik.tu-muenchen.de

Abstract

This paper presents a multi-resolution approach for automatic extraction of roads from digital aerial imagery. Roads are modeled as a network of intersections and links between the intersections. For different context regions, i.e., rural, forest, and urban areas, the model describes relations between background objects, e.g., buildings or trees, and road objects, e.g., road-parts, road-segments, road-links, and intersections. The segmentation of the image into context regions is done by texture analysis. The approach to detect roads is based on the extraction of edges in a high resolution image and the extraction of lines in an image of reduced resolution. Using both resolution levels and explicit knowledge about roads, hypotheses for roadsides are generated. The roadsides are used to construct quadrilaterals representing road-parts and polygons representing intersections. Neighboring road-parts are chained to road-segments. Road-links, i.e., the roads between two intersections, are built by grouping of road-segments and closing of gaps between road-segments. For the road-segments markings give highly reliable evidence. Road-links are constructed using knowledge about context.

Keywords: Road Extraction, Multi-Resolution, Context

1 Introduction

In digital photogrammetry there are operational automatic solutions for geometric tasks, e.g., measurement of fiducials, matching of points for relative orientation and generation of digital surface models. Semantic tasks, like data capture and update of geographic information systems (GIS), still have to be performed manually by a human operator. This is very time consuming and expensive. Therefore, automatic solutions are highly desirable.

Research on the automatic extraction of man-made objects, like buildings or roads, from aerial or satellite imagery has been carried out since the seventies. These days more and more work focuses on automatic image interpretation for GIS (in this paper site-models are also regarded as a kind of GIS). Even though, for GIS update, i.e., verification and acquisition, old GIS data can make object extraction easier, the proposed approach is restricted to object extraction (here: roads) without using prior GIS information (for work in this area see, e.g., (de Gunst 1996, Wiedemann and Mayer 1996)). There are two reasons for this approach: Automatic extraction shows the potential and limits of an extraction scheme much better than GIS-guided extraction and can deepen the understanding of image interpretation. What is more, the extraction of new roads is needed for the update in any case.

A lot of work exists on the extraction of roads. The most common methods are extraction or tracking of lines in small scale imagery and profile matching or detection of roadsides in large scale images. The approaches combine different methods and incorporate additional knowledge, e.g., geometrical constraints, in various ways. A main criterion to distinguish the works is the interaction of a human operator. In semi-automatic schemes an operator selects an initial point and a direction for a road tracking algorithm (McKeown and Denlinger 1988, Vosselman and de Knecht 1995). In (Grün and Li 1996, Merlet and Zerubia 1996) the operator marks a few points of a road segment and an algorithm, e.g. based on dynamic programming, finds the road which connects these points (also in 3D for more than one image). This is advantageous because the path of the road is more constrained, and a more reliable handling of obstacles is possible. A similar approach based on so-called "ziplock" snakes is given in (Neuenschwander et al. 1995). These semi-automatic approaches can be extended to fully automatic operation by means of automatic seed point detection (Zlotnick and Carnine 1993). A fully automatic approach is presented in (Barzohar and Cooper 1996). Stochastic methods are used to find seeds for road extraction. Roads are found by dynamic programming based on a grey level model and on assumptions about the geometry of roads. In (Ruskoné et al. 1994) seed points for the extraction of the road network are centers of elongated regions found by a watershed-based segmentation on a gradient image. Based on the elongated regions and their directions, road segments are extracted using the homogeneity of the road surface. In order to extract the road network geometrical constraints are taken into account, and hypotheses about connections between road segments are checked.

If relations between roads and other objects, like cars, buildings, or trees, are neglected a reliable extraction is often difficult. Background objects can have a strong influence on the characteristics of roads, or at least on the appearance of roads in aerial imagery, e.g., high objects causing occlusions or casting shadows. Therefore, (Strat and Fischler 1995) strongly demand the use of contextual information to guide extraction algorithms for complex scenes. In (Ruskoné et al. 1996) the detection and grouping of cars is used for road extraction in urban areas where profile matching or detection of roadsides would probably fail. (Bordes et al. 1995) analyze the influence of context on the accuracy of roads in a cartographic database.

As distinct characteristics of roads can be detected best at different resolution levels and in different contexts, this paper proposes a multi-resolution extraction scheme integrating context information. The use of context allows to analyze road hypotheses more thoroughly and to deliver less erroneous results. For the generation of hypotheses for roadsides two resolution levels are used. Edges are extracted in a high resolution image and lines in an image of reduced resolution. Using both resolution levels and explicit geometrical and radiometrical knowledge about roads, hypotheses for roadsides are generated and quadrilaterals representing road-parts are constructed from them. Neighboring road-parts are chained, resulting in longer road-segments. Road-parts and road-segments are semantic objects with attributes and methods. The ability of road objects to explicitly represent a specific area also facilitates looking for additional evidence, e.g., road markings, which otherwise is very hard to find. Road-links, i.e., roads between two intersections, and intersections are built from road-parts and road-segments. The attributes and methods of road objects are not exactly the same all over the image but vary due to different contexts. To describe the different appearance of roads three types of so-called context regions are introduced: urban, rural, and forest. For each context region special relations between road objects and background objects are considered (so-called context sketches).

In Section 2, the road model and relations between roads and background objects are described. Section 3 shows the fusion of different resolution levels for the extraction of roadsides and the construction of road objects, along with their attributes and methods. The segmentation of the image into different context regions based on texture analysis, and how the context information is used, is explained in Section 4. Finally, in Section 5 conclusions are given.

2 Model

In order to extract roads it is necessary to have a clear idea about the concept *road*. Hence, a model should comprise explicit knowledge about road width, parallelism of roadsides (geometry), reflectance characteristics (radiometry), network structure (topology), and relations to other objects like buildings or trees (context). The model described below is split into two parts. One part describes characteristics of roads (Sect. 2.1). The other part defines relations to background objects, called context sketches here, and assigns these relations to different context regions (Sect. 2.2). In this way the complex model of the object *road* is divided into more specific sub-models which are adapted to contextual environment (urban, suburb, rural, forest), sensor, and scale (resolution). The sub-models emphasize certain characteristics of the objects and therefore can be regarded as specialized models.

2.1 Model of Road Objects

Roads in Real World: In the real world, the characteristics of roads can be described as consequence of their function for human beings (Stark and Bowyer 1991). Roads make space accessible and therefore exhibit the form of a network connecting all areas inhabited and used by human beings. The denser an area is inhabited and the more intensive it is used, the denser is the road network of the area. Corresponding to their importance, network components are classified into a hierarchy of different categories with different attributes. Roads of national importance, i.e., highways, are much wider than roads accessing agricultural or forest areas. Their relation with the terrain surface differs according to their function and category. Agricultural paths and less important feeder roads are more closely following the natural terrain surface than highways do, the latter ones serving as fast connections between densely settled areas. Roads of different categories therefore differ with respect to minimum curvature radius and maximally allowed slope. As a consequence of greater curvature radius and more gentle slope it happens more often that roads of a greater importance run along embankments, over bridges or in tunnels. On closer inspection, distinct parts of roads can be attributed by the type of surface material, the existence and type of road markings, or the existence of sidewalks and cycle-tracks.

Roads in Aerial Images: The appearance of roads in digital imagery strongly depends on the sensor's spectral sensitivity and its resolution in object space. Usually roads show up with specific grey-values, especially in color and infra-red images. The remainder of this paper is restricted to grey-scale images and only scale dependencies are considered. Images with varying resolution show different characteristics of roads. In images with low resolution, i.e., more than 2 m per pixel, roads mainly appear as lines establishing a more or less dense network. Opposed to this, in images with a higher resolution, i.e., less than 0.50 m, roads are depicted as elongated homogeneous areas with more or less parallel borders and almost constant width. Usually the brightness of roads is different from the surroundings and therefore the roadsides can be extracted using edge detectors. In the higher resolution the attainable geometric accuracy is better, but background objects like cars, trees, and buildings disturb the road extraction to a higher degree.

In a smoothed image — which corresponds to a reduced resolution — lines representing road axes can be extracted in a stable manner even in the presence of those background objects. Figure 1a) displays a bar shaped bright line (= road) with a bright disturbance (= car) on it and its behavior in scale-space for the line extraction model presented in (Steger 1996). In Figure 1b) the position of edge and line points at different scales is plotted against their ideal position.

For small σ the extracted line position will be determined by the bright object, while for large σ it will correspond to the center axis of the road. Therefore, simply by increasing σ one can eliminate the car from the road. It can also be seen that the two edges corresponding to the bright object will vanish along with the flat inflection points on the undisturbed part of the line. As (Mayer and Steger 1996) have shown,

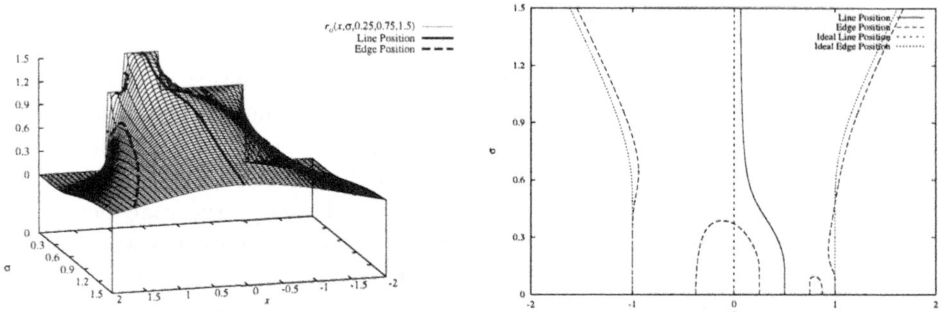

(a) Line and edge position in scale-space (b) Extracted and ideal line and edge position

Fig. 1: Scale-space behavior of a line with a bright disturbance on it.

the appropriate scale for line extraction can be computed, if the width of the line (=road) and of the disturbance (=car) as well as the contrast between background, line, and disturbance is given. Seen from a symbolical point of view, in the larger scale the substructure of the road (the car on the road or also objects like markings) has been eliminated. This can be interpreted as the *abstraction*, i.e. the increase of the level of simplification and emphasis of the road. Abstraction is achieved simply by changing the scale of the object (Mayer and Steger 1996).

Road Model: From the last paragraph follows that fusion of low and high resolution results can contribute to improve the reliability of the road hypotheses. Additionally, details like road markings, which can be recognized at a resolution of less than 0.20 m, can be used as evidence to corroborate the detected road hypotheses. On the one hand, using multiple resolution levels improves robustness of road extraction. On the other hand, it results in the necessity to use different features at each resolution level, and to simultaneously combine all features of all resolution levels into one road model. The semantic network in Figure 2 illustrates the road model condensed from this.

The model is split into three levels, defining different points of view. The *real world* level consists of the objects and their relations on a natural language level. The road of this level consists of intersection and road-link, which connects intersections. The road-link is constructed of short parts, the road-segments. In large scale the road-segment is constructed of road-parts (for the motivation of their use see Sect. 3.2) which in turn comprise the pavement and the markings (solid or dashed). The objects in the *real world* level are connected to the objects in the *geometry and material* level by means of the *concrete* relation which connects *concepts* describing the same object on different levels, i.e., from different points of view. The *geometry and material* level is an intermediate level which represents the three-dimensional shapes of objects as well as their material (Tönjes 1996). This level has the advantage that it represents objects independently from sensor characteristics and viewpoint, which is in contrast to the *image* level.

Road-segments in small scale are linked to a mostly straight bright line in the *image* level. In contrast to this, the pavement of the large scale is linked to the elongated bright area in the *image* level via the elongated flat concrete of asphalt area in the *geometry and material* level. The markings are related to bright lines via colored lines.

Whereas the large scale gives detailed information, the small scale adds global information. If the information of both levels is fused, false hypotheses for roads are eliminated by using the abstract small scale information, while integrating details from the large scale (e.g., the correct width of the roads). With this, the advantages of both scales are merged.

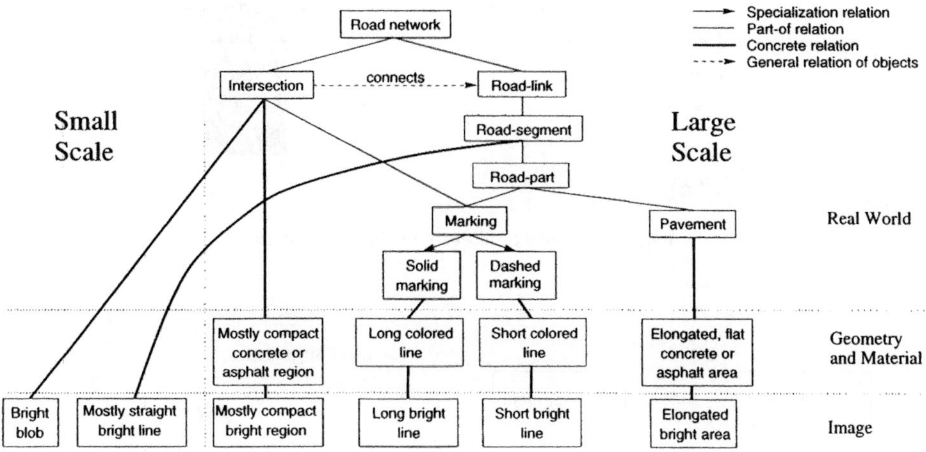

Fig. 2: Road model: different resolutions and points of view

2.2 Context

The road model of the last section comprises knowledge about radiometric, geometric, and topological characteristics. But besides features which refer to the road itself, it is also necessary to consider relations between roads and other objects like buildings, trees, and cars. This is, because these background objects on the one hand support and on the other hand hinder road extraction.

Context Sketches: The notion "context sketch" is introduced to describe typical relations between road objects and background objects. The context sketch *occlusion_shadow* consists of a hypothetical road-part which connects two road-segments and a high object next to the hypothetical road-part. Figure 3a) illustrates an occlusion, and Fig. 3b) a shadow cast onto the road by a high object. Another context sketch describes the relation between road-segments and driveways to agricultural fields (*rural_driveway*). Figure 4a) explains the situation where a driveway causes a gap between two road-segments and only one roadside is found. The context sketch *vehicle_road* describes a dark or bright blob on a road-part or on a road-segment. To be a vehicle the blob must have a certain size and it should not be centered on the road (cf. Fig. 4b). The context sketch *building_driveway_road-segment* models the relations between a road-segment, a driveway (i.e., a small and short road-part), and a building at the end of the driveway. The driveway is often almost perpendicular to the road-segment and the building is in many cases parallel to the road-segment. The relation between roads and parallel objects, like sidewalks and cycle-tracks, is defined by the context sketch *parallel_object.*

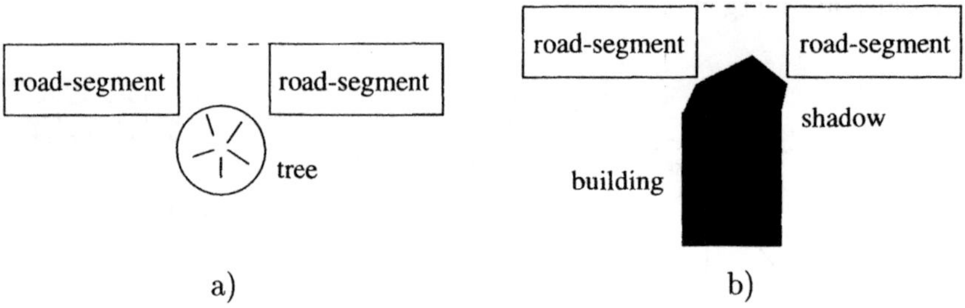

Fig. 3: Context sketch *occlusion_shadow*

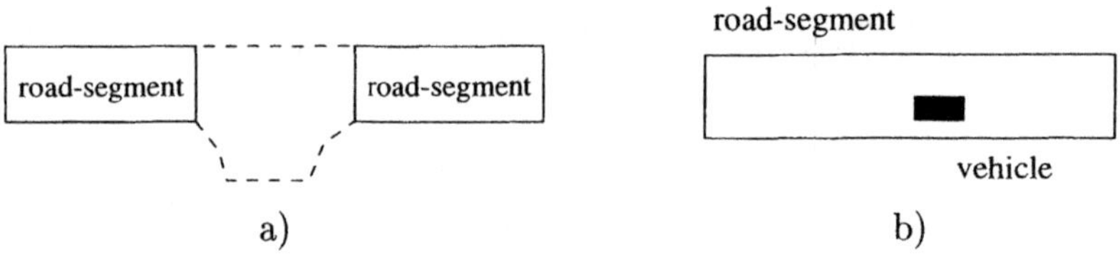

Fig. 4: Context sketches: a) *rural_driveway* b) *vehicle_road*

These basic context sketches are modules that can be aggregated into more complex context sketches depicting, e.g., the interaction between *building_driveway_road-segment* and *occlusion_shadow* or the interaction between *building_driveway_road-segment* and *vehicle_road.*

Context Regions: Not every context sketch always has to be taken into account. This is in contrast to approaches representing the whole scene with one model (Matsuyama and Hwang 1990). The relevance of features and relations depends also on the global context. Roads in urban or suburban areas look quite different and have other relations than roads in rural or forest areas. Therefore, this paper proposes to use different features and relations, i.e., context sketches, not only at multiple resolution levels but also within so-called context regions, for which *suburb_urban*, *forest*, and *open_rural* areas are distinguished here. Information from a GIS or a (pre-)segmentation of the image into these regions provides global a priori information about the characteristic features and their relations. For example, buildings in downtown areas are — in contrast to buildings in rural areas — very close and highly parallel to roads; sidewalks and cycle-tracks are more likely to appear in urban areas; in rural areas single trees and single buildings might hinder extraction, whereas in forest regions mainly shadows and occlusions pose problems. This information about existence or proximity of background objects makes it possible to choose the most appropriate extraction algorithm and to determine the meaning of distinct road-parts and road-segments. The assignment of the context sketches to context regions is shown in Figure 5.

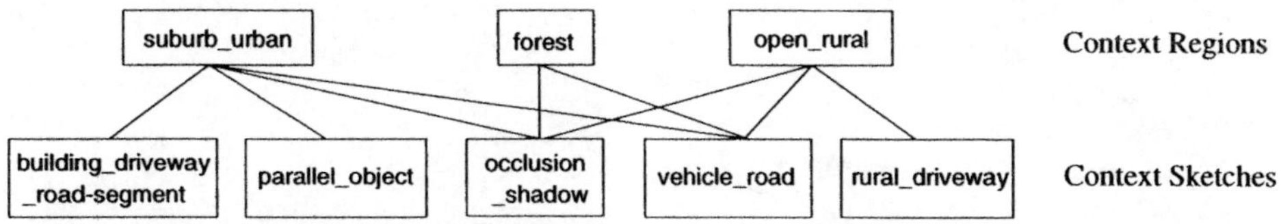

Fig. 5: Context regions and context sketches for roads

3 Strategy

The main principle in the proposed extraction scheme is to guide road extraction by given information. Therefore, the most salient objects are extracted first. Depending on the area where roads are to be extracted, different features are easiest to detect. In urban and forest areas radiometric and geometric knowledge alone is not sufficient due to shadows and occlusions. On the other hand, with a simple model, relying only on attributes of the road itself, good results are expected for rural areas. According to these considerations the road extraction starts in rural areas mainly based on low level algorithms. The intermediate results after this step can then serve to guide road extraction in urban and forest areas. To enable the reasoning in object space (i.e., to use information about the scene in meters and not in pixels), the road extraction is done in an orthophoto. This gives enough information in most cases, since roads are planar objects. According to the considerations explained above, not only images at original resolution but also images at reduced resolution are used. A segmentation of the image into different context regions is achieved by texture analysis in the image of reduced resolution. In a first step, roadsides are extracted using a multi-resolution approach, which integrates edge extraction in a high resolution image and line extraction in a low resolution image (Sect. 3.1). In a second step, road objects, like road-parts, road-segments, and intersections, are constructed from the roadsides (Sect. 3.2). The road objects have attributes and methods to enable a more sophisticated examination of reasons for gaps between road-parts and road-segments. The role of context information for road extraction is shown in Section 4.

3.1 Extraction of Roadsides

Low Resolution: Lines "Low resolution" in this context means that roads are only a few pixels wide and appear as bright or dark lines. Therefore, the resolution considered as "low" depends on the width of the roads in the imagery. If the road width varies widely in an image, e.g., between 4 m (path) and 30 m (highway), more than one "low resolution" level, e.g., one for paths and normal roads and one for highways, would be needed.

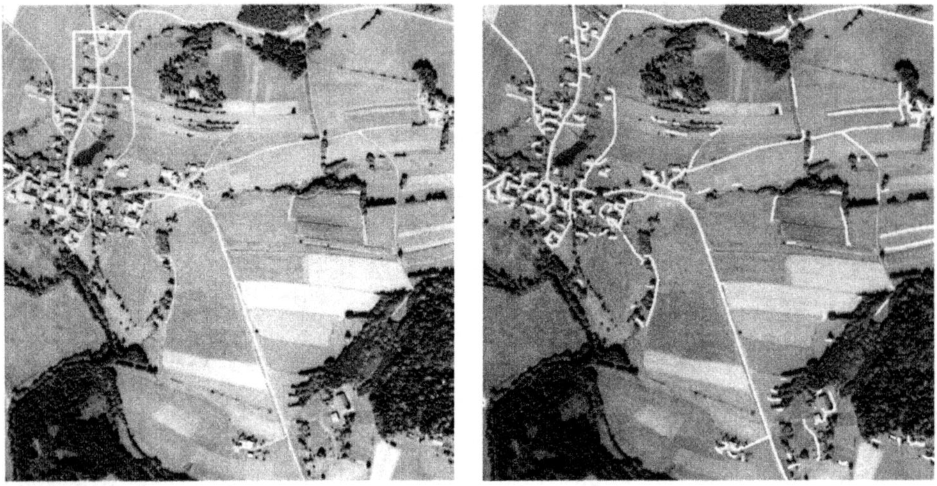

a) b)

Fig. 6: a) Image at low resolution. b) Hypotheses for road axes at low resolution.

Figure 6a) shows a version of reduced, i.e., low resolution of an image. The ground resolution is 2 m. Bright lines are extracted with an approach based on differential geometrical properties of the image function. Points which have a vanishing gradient and a high curvature in the direction perpendicular to the line are considered as line points and linked to contours. The contours are approximated by polygons which are hypotheses for road axes (cf. Fig. 6b).

High Resolution: Edges At high resolution roads are modeled as bright homogeneous areas bordered by parallel edges. Edges are extracted from the original image and approximated by polygons. The polygons are grouped into relations of pairs of parallels, and the area enclosed by the parallels is examined. The result of this process are hypotheses for roadsides at the high resolution level. Details are given in (Steger et al. 1995).

Fusion of Low and High Resolution As motivated in Section 2.1, the results should be fused, to make optimal use of the advantages of both resolution levels. Input for fusion are hypotheses for road axes from low resolution (cf. Sect. 3.1), hypotheses for roadsides from high resolution, and the original edge polygons. The basic strategy is to select the results of both levels supporting each other. Hypotheses for roadsides which enclose a homogeneous area and have a hypothesis for a road axis in-between serve as initial hypotheses. Further roadsides are extracted gradually according to several rules. Figure 7a) shows the input to fusion. The final roadside hypotheses are presented in Figure 7b).

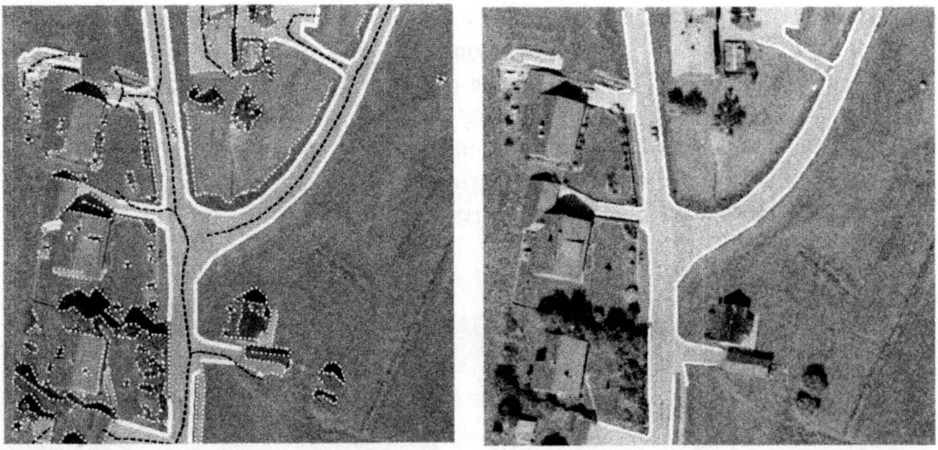

a) b)

Fig. 7: a) Input to the fusion process: hypotheses for road axes (dashed, black), hypotheses for roadsides (solid, white) and other edges (dotted, white). b) Final roadside hypotheses.

3.2 Extraction of Semantic Objects

The results obtained in the last section can be regarded as mid-level segmentation results. This means, that even if they look like roads in Figure 7b), they have not been extracted so far; candidates for roadsides have been found. In order to facilitate further processing, semantically meaningful objects with rich attributes have to be generated.

As shown in Figure 2, the road network consists of *intersections* and *road-links* and can therefore be regarded as a graph. However, it is rarely possible to extract a single connected road-link between two intersections because of occlusions and other problems. Therefore, the concept *road-segment* is used to describe a single, connected piece of road detectable in the image. At low resolution this concept corresponds directly to the bright lines extracted by the method described in Section 3.1. At high resolution a road-segment can be described by its enclosing polygon. However, it is quite difficult to generate this polygon directly in a topologically correct way. Therefore, *road-parts* are introduced as intermediate concepts. They have a simple geometric description, namely the minimal quadrilateral built by two parallel and overlapping roadsides. Road-segments can then simply be created by aggregating adjacent road-parts. Similarly, at low resolutions intersections are junctions of more than two lines. Conversely, at high resolution they are regions described by their enclosing polygon.

Extraction of road-parts: The first step to generate the different types of road objects is to generate road-parts from the results obtained in Section 3.1. This is done by splitting the parallels into non-overlapping quadrilaterals by projecting each point of the left and right parallel polygons perpendicularly onto the other polygon. Each road-part is described by the four points of the quadrilateral. Furthermore, there are functions in the corresponding class to calculate the center axis of each road-part, and to convert the road-part into a region in the image. The latter point facilitates the selective extraction of features on the road-part, e.g., road markings. Figure 8a) shows the extracted road-parts.

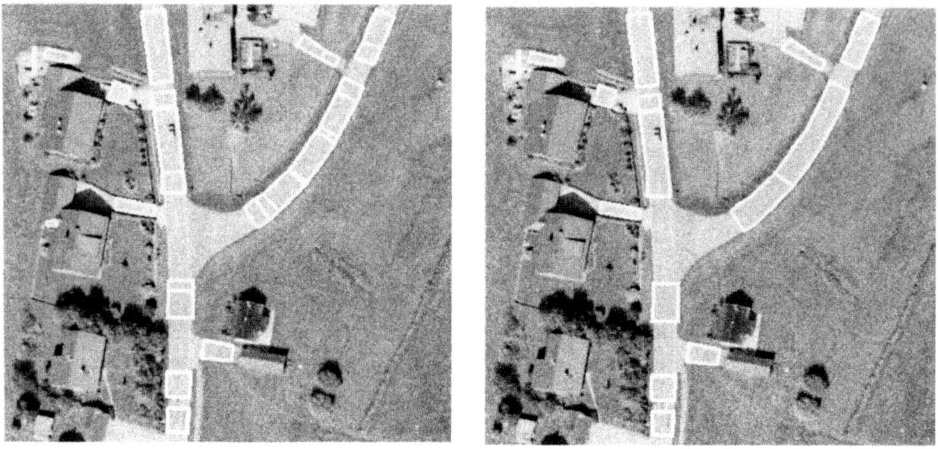

a) b)

Fig. 8: a) Road-parts b) Road-segments

Aggregation of road-parts into road-segments: When the road-parts have been extracted, it is straightforward to group them into road-segments by collecting all adjacent road-parts. The road-segments inherit much of their attributes and functionality from their constituting road-parts, e.g., the abilities to generate their center axis, and to be converted into regions. Figure 8b) shows the road-segments generated from the road-parts in Figure 8a). It is apparent that most of the roads are correctly extracted. It can also be seen that sometimes the generation of road-parts fails to cover the entire area between parallels because at some areas no quadrilaterals can be generated, especially in highly curved areas. However, this can easily be corrected. The road-segments thus created can be aggregated into road-links by grouping processes. As this is context-dependent, it will be described in Section 4.

Extraction of intersections. In Figure 2 intersections are areas in the image where more than two road-links meet. At low resolution they are points where three or more lines meet (junctions). At high resolution they are complicated objects defined by their enclosing polygon. Candidates for this polygon are the roadsides remaining after the above steps which border a road-like region, but are not parallel. However, to find out which roadsides belong to a particular intersection is not trivial. For this task information from the low resolution level is used. Hypotheses for intersections are junctions from the road network graph. Since the line detector fails to extract junctions under predictable circumstances, the algorithm completes the graph by adding junctions at places where a line end is very close and almost perpendicular to another line. The lines adjacent to the junction are then examined whether they cross road-segments at high resolution. This procedure yields road-segment candidates for a particular intersection. Since the lines adjacent to a junction can be quite long, and cross several road-segments at high resolution, the closest road-segments on each line are selected as belonging to the intersection. They define a search area for the enclosing polygon of the intersection. Polygons in this area are then aggregated to the enclosing polygon of the intersection by tracing them from the end points of the adjacent road-segments, and if necessary, by closing gaps where no roadsides or edges were found. Figure 9a) shows the road-segments and the network graph derived from low resolution, while Figure 9b) displays the extracted intersection polygons. As can be seen, this procedure works quite well for the case of junctions where three road-segments meet. When four or more road-segments form the intersection the procedure has to be slightly modified, since the line extractor will usually extract two closely adjacent junctions of three lines in this case. Like road-segments, intersections have a rich set of attributes (e.g., the adjacent road-segments) and functions (e.g., they can be converted into discrete regions for low level processing).

4 Role of Context

4.1 Context Regions

As described in Section 2.2, context regions provide additional global information enabling a more efficient use of knowledge about roads. To construct the context regions, available GIS data can be used. As the GIS data can be outdated, it should be compared with information from images, e.g. from a segmentation. If the borders of GIS regions and the segmented regions do not differ too much, the GIS borders are supposed to be correct. Otherwise the borders of context regions are taken from the segmentation. In cases where there is no GIS data available, the context regions can be obtained from the segmentation alone. A segmentation of the aerial image into the context regions *open_rural, forest,* and *suburb_urban* area can be achieved by texture analysis. The texture features which can be used depend on the resolution of the image. The main textural characteristic of *open_rural* area in low resolution is homogeneity. Besides edges at the borders of fields or meadows as well as next to roads, there is almost no contrast in this context region. On the other hand in *forest* and *suburb_urban* areas the texture energy is much higher. Therefore, the

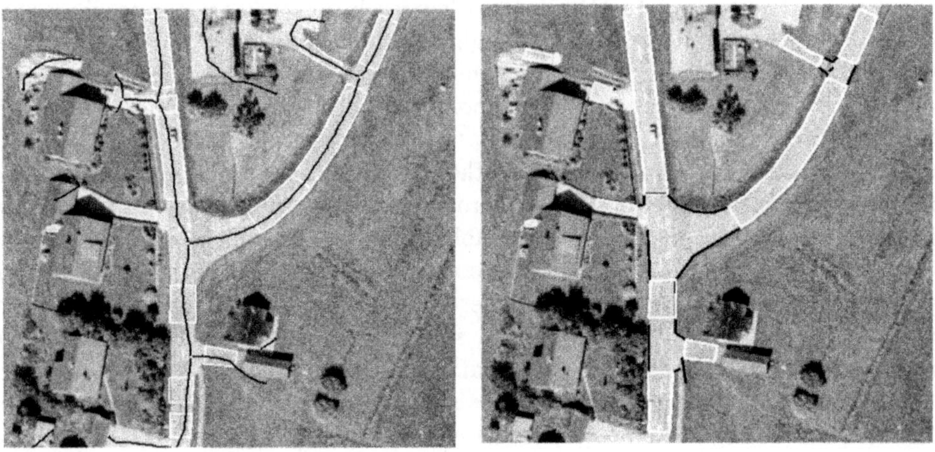

a) b)

Fig. 9: a) Road-segments (white), lines (black) and junctions (white points). b) Polygons of intersection areas (black) and road-segments (white).

segmentation of *open_rural* area is not too difficult using the texture energy derived from images at a resolution of 4 m filtered with masks proposed in (Laws 1980). However, at this resolution a reliable distinction between *forest* and *suburb_urban* area by texture energy alone is not possible and has to be done by other criteria and/or at higher resolution. A quite simple but efficient way is thresholding of the original grey-values and application of morphological operations on the results of the thresholding. Figure 10 shows results of the segmentation.

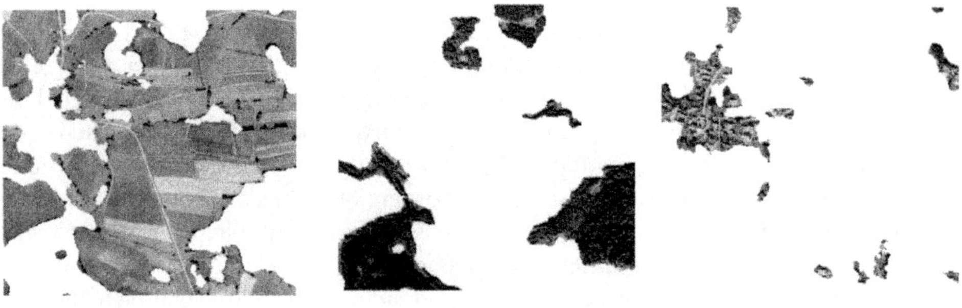

a) b) c)

Fig. 10: Segmentation at low resolution: a) *open_rural* b) *forest* c) *suburb_urban*

4.2 Road Extraction in *open_rural* Area

To extract those parts of the road network which are in the *open_rural* context region, all road-parts, road-segments and intersections in *open_rural* area are selected. The goal is to close the gaps between the road-segments and to generate road-links, i.e., long road-segments connecting two intersections. Figure 11 shows an example for the input data to this step.

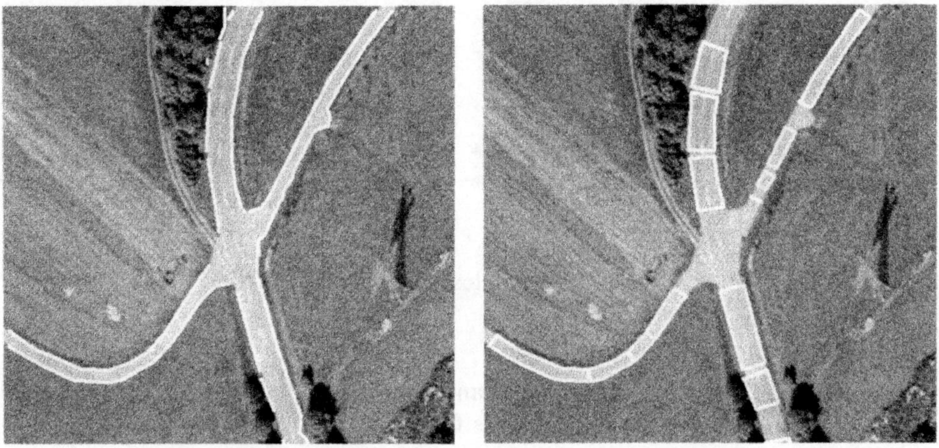

a) b)

Fig. 11: a) Roadsides b) Road-segments

In *open_rural* area there are two main reasons for gaps: First, small gaps caused by the preceding steps (extraction of roadsides, construction of road-parts, etc.) and second, longer gaps caused by local disturbances, like single trees next to the road or branching driveways to the fields. Gaps of the first kind usually do not exceed more than a few meters, and can be closed if the road-segments are connected by an extracted roadside at one side. Gaps of the second kind can be handled using the context sketches *occlusion_shadow* and *rural_driveway*. If a gap can be explained by a context sketch, a road-part is constructed to connect the two road-segments (cf. Fig. 12a).

After this step, all road-segments ideally should connect two intersections, i.e., they are road-links, or they should end at the border of the *open_rural* area or the image. Road-segments with dead ends inside the *open_rural* area have to be checked whether they are really roads. To accept such road-segments as road-links, an important object like a building has to be close to the dead end.

Additional and very strong evidence for a road is gained from the extraction of road markings which appear as weak bright lines in the high resolution image. They are found using the line extractor described in Section 3.1 (cf. Fig. 12b). Of course, reasonable results are possible only because the line extraction is restricted to the

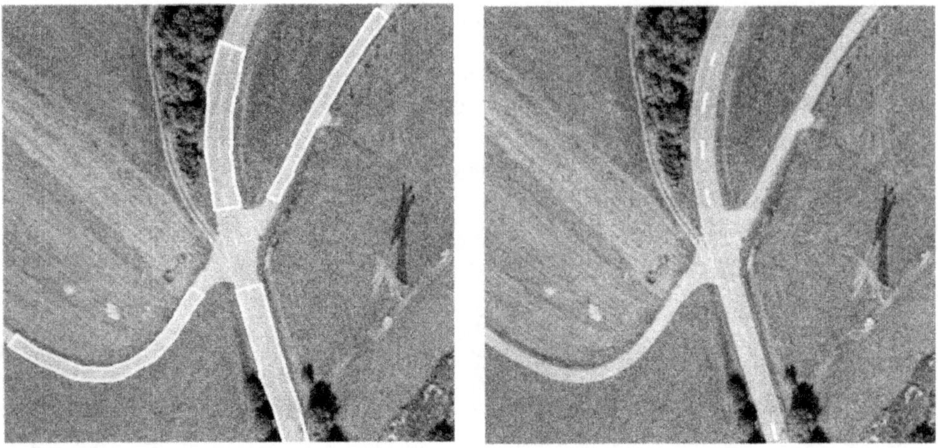

a) b)

Fig. 12: a) Road-segments after closing gaps of second-kind b) Markings

hypothetic road area: Faint markings can only be found using very low thresholds, which would result in a plethora of lines for other parts of the image.

4.3 Outlook: Road Extraction in *suburb_urban* and *forest* Areas

For the extraction of roads in *suburb_urban* and *forest* areas road-segments of the *open_rural* area ending at the border to the other context regions can be used as reliable starting points for a road tracking into the *suburb_urban* and *forest* areas. Hypotheses for paths which connect the starting points are provided by the line extraction at low resolution. The road network is propagated from the border to the center of the *suburb_urban* and *forest* areas. In order to achieve good results, it will be necessary to integrate more context information, especially about high objects and shadows. This additional information will be employed by means of context sketches.

5 Summary

In this paper a multi-resolution approach for road extraction from aerial imagery is presented which makes use of local (sketch) and global (region) context information. The road model describes objects of the road network and relations between these objects and other objects. More specifically, roadsides are extracted by fusing the results of line extraction at low resolution and edge extraction at high resolution. From the result road-parts, road-segments, and intersections are constructed. These road objects have a clear semantics as well as attributes and methods for further reasoning. Integration of context regions restricts possible context sketches (i.e., relations to background objects) and makes the use of knowledge more effective.

Acknowledgments

This work is embedded in the joint research project *Semantic Modeling and Extraction of Spatial Objects from Images and Maps* and is funded by the Deutsche Forschungsgemeinschaft (DFG) under grant no. Eb 74–8/2.

References

Barzohar M., D. Cooper (1996) *Automatic Finding of Main Roads in Aerial Images by Using Geometric–Stochastic Models and Estimation*, IEEE Transactions on Pattern Analysis and Machine Intelligence, Vol. 18, No. 7, pp. 707–721.

Bordes G., G. Giraudon, O. Jamet (1995) *Road Extraction Guided by a Cartographic Database: Creating a Strategy*, International Conference on Computer Vision, Workshop on Context-Based Vision, pp. 30–37.

de Gunst M. (1996) *Knowledge-Based Interpretation of Aerial Images for Updating of Road Maps*, Netherlands Geodetic Commission, Publications on Geodesy, New Series 44, Delft, The Netherlands.

Grün A., H. Li (1996) *Linear Feature Extraction with LSB-Snakes from Multiple Images*, International Archives of Photogrammetry and Remote Sensing, Vol. (31) B3/III, pp. 266–272.

Laws K. (1980) *Texture Image Segmentation*, PhD thesis, University of Southern California, Deptartement of Engineering, Los Angeles.

Matsuyama T., V.-S. Hwang (1990) *SIGMA - A Knowledge-Based Aerial Image Understanding System*, Plenum Press, New York.

Mayer H., C. Steger (1996) *A New Approach for Line Extraction and its Integration in a Multi-Scale, Multi-Abstraction-Level Road Extraction System*, IAPR TC-7 Workshop: Mapping Buildings, Roads and other Man-Made Structures from Images, Oldenbourg Verlag, Wien, Austria, pp. 331–348.

McKeown D., J. Denlinger (1988) *Cooperative Methods For Road Tracking In Aerial Imagery*, Computer Vision and Pattern Recognition, pp. 662–672.

Merlet N., J. Zerubia (1996) *New Prospects in Line Detection by Dynamic Programmming*, IEEE Transactions on Pattern Analysis and Machine Intelligence, Vol. 18, No. 4, pp. 426–431.

Neuenschwander W., P. Fua, G. Székely, O. Kübler (1995) *From Ziplock Snakes to Velcro^{TM} Surfaces*, Automatic Extraction of Man-Made Objects from Aerial and Space Images, Birkhäuser Verlag, Basel, Switzerland, pp. 105–114.

Ruskoné R., L. Guigues, S. Airault, O. Jamet (1996) *Vehicle Detection on Aerial Images: A Structural Approach*, 13th International Conference on Pattern Recognition, Vol. III, pp. 900–903.

Ruskoné R., S. Airault, O. Jamet (1994) *A Road Extraction System Using the Connectivity Properties of the Network*, Zeitschrift für Photogrammetrie und Fernerkundung, Vol. 5/94, pp. 174–180.

Stark L., K. Bowyer (1991) *Achieving Generalized Object Recognition through Reasoning about Association of Function to Structure*, IEEE Transactions on Pattern Analysis and Machine Intelligence, Vol. 13, No. 10, pp. 1097–1104.

Steger C. (1996) *An Unbiased Detector of Curvilinear Structures*, Technical Report FGBV-96-03, Forschungsgruppe Bildverstehen (FG BV), Informatik IX, Technische Universität München.

Steger C., C. Glock, W. Eckstein, H. Mayer, B. Radig (1995) *Model-Based Road Extraction from Images*, Automatic Extraction of Man-Made Objects from Aerial and Space Images, Birkhäuser Verlag, Basel, Switzerland, pp. 275–284.

Strat T., M. Fischler (1995) *The Role of Context in Computer Vision*, International Conference on Computer Vision, Workshop on Context-Based Vision, pp. 2–12.

Tönjes R. (1996) *Knowledge Based Modelling of Landscapes*, International Archives of Photogrammetry and Remote Sensing, Vol. (31) B3/III, pp. 868–873.

Vosselman G., J. de Knecht (1995) *Road Tracing by Profile Matching and Kalman Filtering*, Automatic Extraction of Man-Made Objects from Aerial and Space Images, Birkhäuser Verlag, Basel, Switzerland, pp. 265–274.

Wiedemann C., H. Mayer (1996) *Automatic Verification of Roads in Digital Images Using Profiles*, Mustererkennung 1996, Springer-Verlag, Berlin, pp. 609–618.

Zlotnick A., P. Carnine (1993) *Finding Road Seeds in Aerial Images*, Computer Vision, Graphics, and Image Processing: Image Understanding, Vol. 57, pp. 243–260.

Buildings

Semanitic Modeling for the Acquisition of Topografic
Information from Images and Maps
ed. by W. Förstner and L. Plümer

Integration of 2D and 3D Reasoning for Building Reconstruction Using a Generic Hierarchical Model

André Fischer[1] and Thomas H. Kolbe[2] and Felicitas Lang[3]

Institut für Informatik I[1]/III[2], Institut für Photogrammetrie[3], Universität Bonn
e-mail: andre@cs.uni-bonn.de, tk@cs.uni-bonn.de, Felicitas.Lang@ipb.uni-bonn.de

Keywords: building modeling, multi image correspondence analysis, mid-level feature aggregates, aspect hierarchies, constraint logic programming.

Abstract

We propose a model-based approach to automated 3D extraction of buildings from aerial images. The semantics of the concept *building* is used to control and to evaluate building extraction in all stages of the process. The semantics is encoded by means of generic 3D object modeling, which describes thematic and geometric constraints on the spatial structure of buildings, and of 2D image modeling, which integrates sensor and illumination modeling to describe the appearance of buildings specific for the given aerial imagery. 3D object and 2D image modeling are tightly coupled within a multi-layered framework which contains an *is-part-of*-hierarchy of 3D building parts and their corresponding image descriptions. The overall strategy follows the paradigm of hypotheses generation and verification and combines bottom-up and top-down processes. Due to the explicit representation of well defined processing states in terms of model-based 2D and 3D descriptions at all levels of modeling and data aggregation our approach reveals a great potential for a reliable building extraction.

1 Introduction

Due to the fact that more than about 50% of the world population live in urban or suburban environments the automation of 3D building extraction is an issue of high importance and shows an increasing need for various applications including geo-information systems, town planning or environmental related investigations.

Aerial images contain on the one hand a certain amount of information not relevant for the given task of building extraction like vegetation, cars and building details. On the other hand there is a loss of relevant information due to occlusions, low contrast or disadvantagous perspective. To compensate for these properties of image data as well as for being able to handle the overwhelming complexity of building types and building structures, a promising concept of automated building extraction from aerial images must incorporate a sufficiently complete model of the objects of interest and their relations within the whole process of image interpretation and object reconstruction (cf. Suetens et al. 1992).

Such a strong and complete modeling approach seems not to be available up to now as the acquisition of spatial data for geo-information systems etc. today still is mainly done by human operators. Therefore, the representation of domain specific semantics is still a crucial subject of discussion and research in 3D building extraction.

We propose a model-based approach to automated 3D extraction of buildings from aerial images. The semantics of the concept *building* is used to control and to evaluate building extraction in all stages of the process. It is encoded by means of a generic 3D object model, which describes thematic and geometric constraints on the spatial appearances of buildings, and a 2D image model, which integrates sensor and illumination modeling to describe the projective appearances of buildings specific for the given aerial imagery.

1.1 Related work

Related work on 3D building extraction — or in general on 3D scene reconstruction — reveals different modeling schemes. Polyhedral models show a long tradition as approximative object descriptions (e.g. Clowes 1971, Huffman 1971, Waltz 1975, Sugihara 1986, Kanatani 1990, Heyden 1994). Obviously polyhedral descriptions are too general for the use within 3D building extraction and therefore move the burden of building modeling on additional representation schemes to represent and organize domain specific heuristics and constraints like in the MOSAIC system (cf. Herman and Kanade 1987) or in the approach of Braun 1994. Parameterized models are restricted to describe the most common building types in the sense of prototypes (cf. McKeown 1990, Quam and Strat 1991, Lang and Förstner 1996a,Lin et al. 1994a, Lin et al. 1994b, Lin et al. 1995) but show a lack to represent variations and combinations of their shapes as well as other relations. Prismatic models can describe arbitrary complex polygonal ground plans of buildings, but reveal the strong restriction to buildings with only flat roofs (cf. Weidner and Förstner 1995, Weidner 1996). CAD models are used to describe objects with fixed geometry and topology in object *recognition* tasks, especially for controlling industrial processes (cf. Hansen and Henderson 1993, Flynn and Jain 1991, Ikeuchi and Flynn 1995, Munkelt 1995). The use of CAD models in building extraction is therefore restricted to the identification of a priori known buildings (cf. Sester and Förstner 1989, Schickler 1993, Huertas et al. 1995). Generic modeling approaches promise on the one hand the greatest modeling power, but on the other hand demand effective constraints and heuristics to restrict modeling to building specific shapes. Fua and Hanson 1987 employ simple box-type primitives but propose an explicit representation of legal primitive combinations to more complex building aggregates. The approaches of Dickinson et al. 1992 and Bergevin and Levine 1993 are from outstanding importance due to the integration of 3D generic object models and an explicit modeling of 2D projective object appearances within a recognition-by-components strategy (cf. Biederman 1987). Both approaches employ volumetric primitives instead of simple box-types but neglect the description of elaborated schemes for domain dependent primitive combinations. Bignone et al. 1996 propose a generic roof model which assumes planar roof surfaces. Extracted 3D roof patches are grouped by an overall optimization according to the simplicity, compactness and completeness of the resulting roof shape. To complete the building shape vertical walls are assumed. This approach shows impressive results on some test data but obviously shows no explicit modeling of building types and building specific aggregation schemes. Groups of planar 3D patches optimized according to the criteria of simplicity, compactness and completeness are not necessarily real roof shapes.

1.2 Overview

In Braun et al. 1995 we proposed in detail the concepts and processes which have to be taken into account for a sufficient complete modeling framework for 3D building extraction. This modeling framework integrates interrelations between image data and model descriptions at different aggregation levels and in terms of corresponding 3D object and 2D projective object descriptions. Within this paper now we present a strategy for a well defined path from the unstructured image data to the model-based and highly structured 3D reconstruction of buildings.

The overall strategy follows the paradigm of hypotheses generation and verification and combines bottom-up (data-driven) and top-down (model-driven) processes. Domain knowledge constraints even the early stages of hypotheses generation due to an elaborated *is-part-of*-hierarchy of 3D building parts and their corresponding projective descriptions. The reconstruction process is carried out already for local 2D feature aggregates to allow an early domain specific classification as 3D local building feature aggregates. A step-wise and strongly model-driven aggregation process combines 3D local building feature aggregates to well defined parameterized 3D building parts and then to more complex 3D building aggregates. The resulting complex 3D building hypotheses and their components are back projected into the images to allow a component-based and robust hypothesis verification applying constraint solving techniques (cf. Kolbe et al. 1996).

Due to the explicit representation of well defined processing states in terms of model-based 2D and 3D descriptions at all levels of modeling and data aggregation our approach reveals a great potential for a reliable building extraction.

2 Concept

In this section we present the proposed building model and discuss its implications on the developed strategy.

2.1 Models

For coping with the complexity of natural scenes we propose an application specific modeling of the domain *buildings*. The general concept has been presented in Braun et al. 1995. It contains a close interaction of bottom-up and top-down strategies

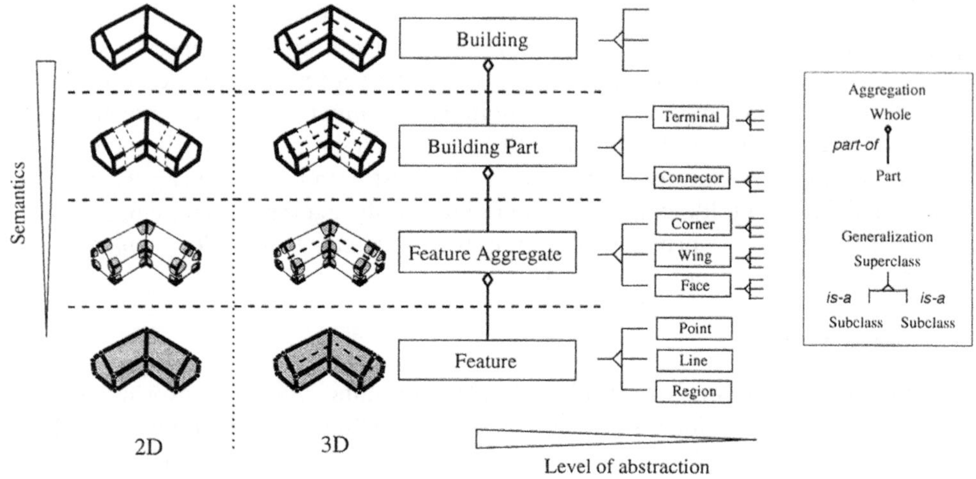

Fig. 1: Building model: The different semantic levels of the part-of hierarchy are shown in vertical direction, the different levels of abstraction of the is-a hierarchy in horizontal direction, which is only shown for the 3D-model. The 2D-image model describes the expected appearance of the building in the different levels of the part-of hierarchy, which is indicated by not showing the hidden lines.

which is fundamental for interpretation tasks dealing with complex image data and models.

A crucial aspect of the approach is the explicit separation of object, sensor and image model in order to reach parsimony in modeling. We start with modeling the 3D-objects leading to the object model. The sensor and the illumination model will transfer many of the concepts from the object to the image model describing the expected appearance of the objects. We now specify these models more detailed than in Braun et al. 1995 and explain the decisive role of the semantics.

Object Model Buildings reveal a high variability in structure which suggests to represent them as an aggregation of several simple building parts. This enables to cope with the problems caused by occlusions, low contrast, noise and disturbances. We therefore propose a multi-level *part-of hierarchy* (cf. figure 1). It reflects different levels of the envisaged semantic abstraction. The primitives of each aggregation level are specialized by an *is-a hierarchy* into subclasses.

Each primitive is described by its semantics, its geometry and possibly its physical properties. Its class membership and its relations to other primitives formally describes its semantics. The geometry is described by pose and form parameters.

We at the moment employ four semantic levels for modeling complex buildings, which seems to be sufficient for a large class of buildings.

The first level (*feature level*) contains features F, namely attributed *points* P , *lines* L and *regions* R. Attributes for lines and regions, for instance, are the orientation classifications horizontal (h), oblique (o) and vertical (v). Regions have an additional attribute describing its role: valid values among others are wall, roof and floor. In general the set of parameters is divided into positional parameters on one hand, describing position and orientation, and form parameters on the other hand like width, height and length.

The second level (*feature aggregate level*) contains feature aggregates A which are induced by points, lines and regions, and contain all their direct neighbors. Each aggregate is defined by a feature graph, given by a set $F = \{f_1, \ldots, f_k\}$ of features and adjacency relations $R \subset F \times F$. A *Corner* C, for instance, contains one point and all its adjacent lines and regions (cf. fig. 2).

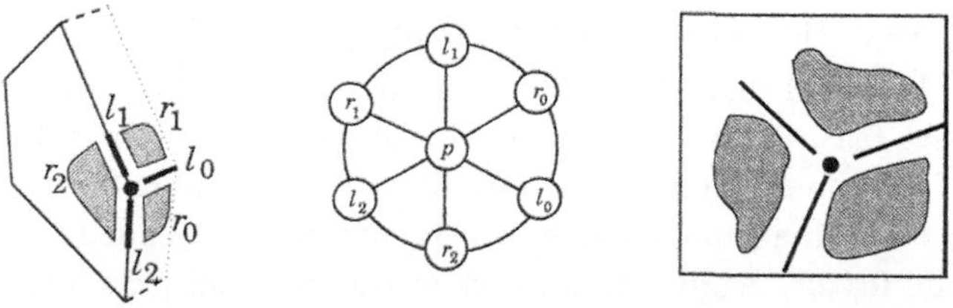

Fig. 2: A corner is a feature neighborhood of a point (left). Drawn as graph the arcs express the adjacency relation (mid). Assuming no occlusions and disturbances its expected appearance in the image reveals the same neighborhood relations between the corresponding 2D-features (right).

The third level (*building part level*) contains building parts P. Currently, they are defined as corner graphs given by a set $C = \{c_1, \ldots, c_n\}$ of corners and adjacency relations $R \subset C \times C$. They are parameterized volumetric objects. Each building part has at least one so called *plug face* which is used for connecting building primitives to each other. We discriminate *terminals* having exactly one plug face and *connectors* with two or more plug faces (cf. figure 3).

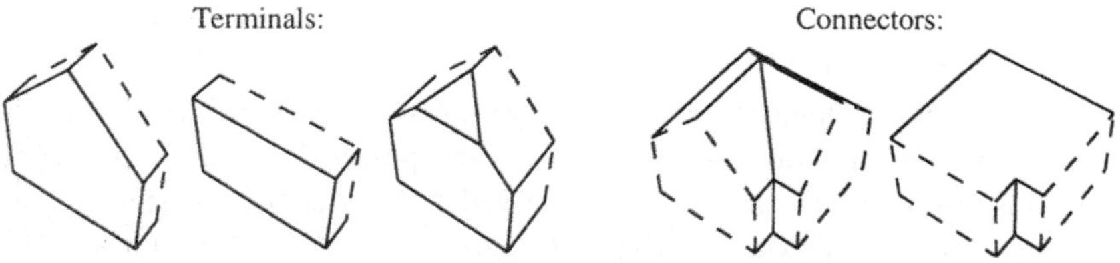

Fig. 3: Some examples of building parts. Plug faces, which are used to connect them, are drawn dashed.

The fourth level (*building level*) contains complete buildings. Buildings are defined as graphs with building parts as nodes, the arcs representing pairs of building parts connected by corresponding plug faces. Thus, the most simple building consists of two connected terminals.

Image Model The 2D image model describes the expected appearance of the building at the same levels of aggregation as the corresponding 3D-structures.[1] This guarantees coherence of the representation of 2D- and 3D-primitives being the prerequisite for all processes of 3D-reconstruction, aggregation, indexing and matching.

The image model on one hand contains all properties which are invariant under projection and can be used as constraints. This especially holds for all thematic labels, neighborhood and geometric relations. E. g. a 2D-corner, being a point induced image aggregate may be labeled a gable corner, the neighborhood relation between two regions of a roof in general can be expected to be transferred to a pair of image region, whereas – assuming weak perspective – parallel 3D-roof lines map to parallel 2D-roof lines. Constraints of the higher aggregation levels are transferred to the primitives of the lower ones, if necessary.

On the other hand all disturbing effects resulting from the illumination, the projection and the feature extraction need to be modeled. This also holds for the appearance of the model internal but not directly visible plug faces. These manifest themselves in local image aggregates, which contain open ended lines and faces, as it occurs in the feature aggregate level or in the building part level (cf. fig. 1 and 3).

The image model will be used for both, the multi view reconstruction of 3D-corners and the 2D-verification of hypotheses of building parts or of complete buildings.

Observations and templates We now are able to describe the task of linking the semantic model with the image data in detail.

As mentioned above, each primitive in the aggregation hierarchy is described by its semantics and its geometry possibly including physical properties. The semantics is manifested in the mutual relations between the different primitives or aggregates, which need to reflect their relations in reality. Semantic modeling therefore consists of describing the models of the primitives — either 3D or 2D — in terms of constraints on their attributes and relations, either on the same or on another aggregation level.

Image interpretation can be seen as a level-by-level goal driven matching process of the original sensor data with the model.

Original sensor data themselves do *not* contain any semantic information. They however provide original or derived geometric and physical measures and implicitely show certain relations. These have to be derived and organized in the same structure as the envisaged class, also in case they are aggregated. An soon as they are linked to primitives of the model by a matching process (cf. below), they result in *observations* which have a meaning and can be used in later steps of the analysis.

As at intermediate levels we obviously obtain only partially instantiated primitives we distinguish two types of instances of the primitives, templates and observations (cf. fig 4):

[1] Actually the image model contains the raster image as the lowest, say 0th level, from which the image features are extracted. This lowest level is not shown as we do not explicitely refer to it.

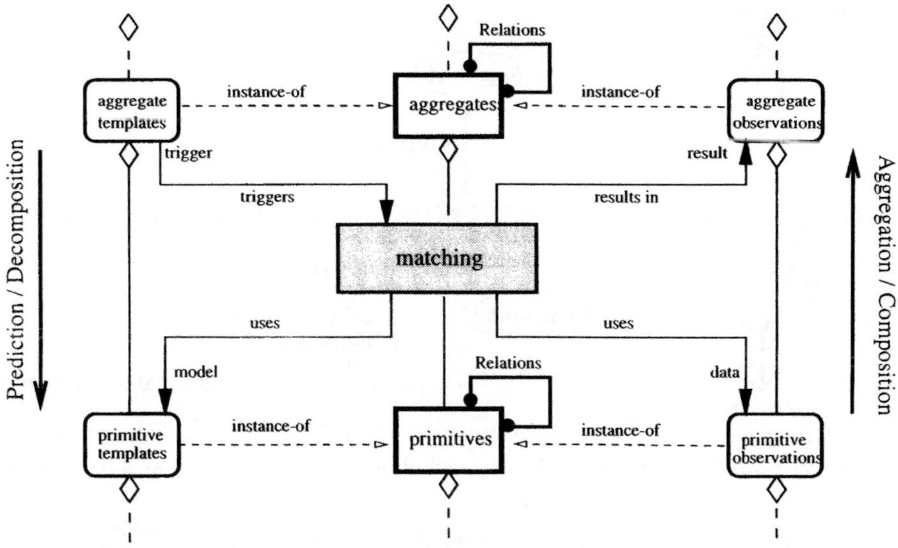

Fig. 4: Templates and observations as partially and fully instantiated primitives. Observations result from image data by a matching process.

1. *Templates.* Templates are partially instantiated classes.

   Templates carry the semantics, which is given them during the modeling process. These are constraints on relations to templates on the same level or on the lower level. Uninstantiated attributes or relations in our context primarily refer to geometric properties, but also to labels of parts of the aggregate, thus templates on the lower level.

2. *Observations.* Observations are fully instantiated primitives.

   Observations obtain their semantics from the template through the matching process, especially by instantiation of mutual relations and relations to primitive observations. Observations can be seen as interpreted image data at a certain level of the aggregation hierarchy.

2.2 Strategy

We employ a combined, data-driven and model-driven strategy integrated in a matching process (cf. Fig. 4). It always links two levels within the *part-of*-hierarchy: Corner reconstruction links the feature and the feature aggregate level, reconstructing building parts or buildings links the feature aggregate level with the building part or the building level, whereas the verification step links the feature level with the building part or the building level.
The matching consists of four steps:

1. *Trigger:* The matching is triggered by a selected template at the higher level e. g. a corner, thus initiated by the semantic model. It can be seen as a set of

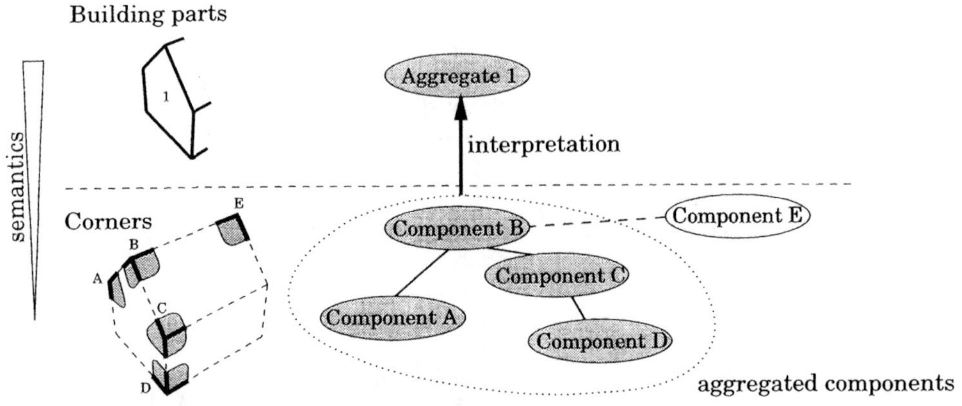

Fig. 5: Grouping and aggregation: components A to E being corners are grouped. Matching leads to a building part as aggregate (excluding E) and labeling it as a terminal gable front.

goals. They have the same structure, e. g. $corner(type, geometry)$, allowing to handle them independently on the specific instance.

2. *Prediction:* The selected goals are used to predict a set of primitive templates on the lower level. They also reveal the same structure and in addition are linked by a set of compatibility relations, which are given by the aggregate template.

3. *Grouping:* Based on the predicted components the primitive observations on the lower level, which play the *role of the data* in the matching process, are grouped. Groups thus are *candidates* for higher level aggregates. E. g. when looking for corners, point induced image structures are established. If strong positional knowledge is available, e. g. when verifying buildings or building parts, grouping consists of finding equivalent primitives, e. g. collinear straight line segments. The data structure of the groups is given by the aggregate template. However, *no final interpretation* takes place as grouping not necessarily checks the mutual relations between the primitives simultaneously.

4. *Aggregation:* The aggregate template is now matched to the found groups, establishing information at the higher level of the hierarchy. In this step all constraints on the geometry, the radiometry and possibly on the semantics of the primitives at the lower level are used. This way primitives contained in the groups may be excluded, cf. Fig. 5, where primitive E is grouped but not aggregated. The matching transfers the semantics of the templates, being part of the model, to the grouped data. This way the groups are interpreted, leading to aggregates which have a specific meaning within the model, e. g. labeling a corner as a gable corner. Using the *part-of* hierarchy, also missing semantic labels for the primitives on the lower level of the hierarchy are established,

indexing them into one of the prespecified classes, e. g. labeling a horizontal 3D-line segment of a vertex as gable point line(h).

2.3 The Procedure

The complete process can be described in the following way: Our input data are given as digital raster images with multiple overlap. Further information about the aerial image flight like exterior and interior camera orientation and time stamp are used. The starting point of our analysis is the extraction of a polymorphic image description consisting of points P^{2D}, lines L^{2D} and regions R^{2D} and their mutual relations (cf. Förstner 1994, Fuchs and Förstner 1995). It allows to derive point, line and region neighborhood aggregates A^{2D}, where vertices V^{2D} are the most promising ones for starting our analysis.

Our procedure consists of three main tasks, which are executed subsequently (cf. Fig. 6):

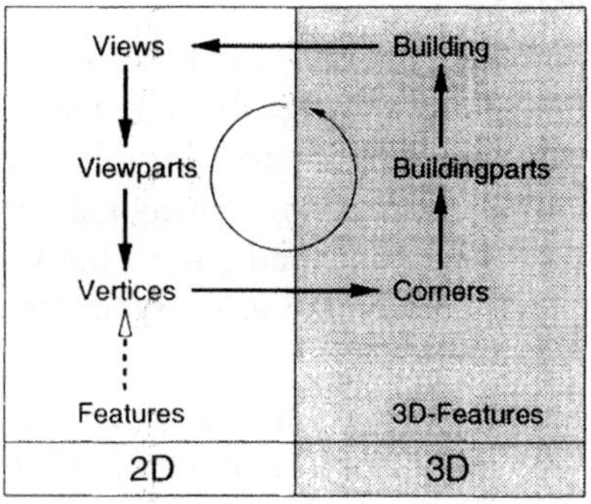

Fig. 6: Information transfer within the whole process. The close integration of 2D and 3D reasoning is performed by an iteration loop. The dashed arrow marks the initialization step. 3D-reconstruction, generation and verification of hypotheses for building parts and buildings is repeated. Verification is based on generated views using matching on the feature level.

1) **reconstruction of 3D corners:** Since the expressiveness of 3D data is higher than that of 2D data, we aim at an early transition from 2D to 3D in the reconstruction process, this way reducing the overall number of future hypotheses. This is done by finding feature groups, namely vertices in the images, which may correspond to building corners. Hypothesized 2D corners of different images are matched, leading to 3D corners.

2) **generation of building hypotheses:** The 3D corners now are taken to find building parts, which can explain these corners. When there is needed more than one building part to explain the 3D observations, the building parts are combined to a building. In this step some components and parameters of the building parts may stay undetermined.

3) **verification of building hypotheses:** For verification the 3D building templates are projected back into the different images, resulting in 2D views for buildings, or views of building parts, (cf. Fig. 1). The predicted feature templates and their interrelationships are then matched with the extracted image features and their relations. The matching possibly determines free parameters of the template.

The successful sequence of matching steps results in an iteratively improved gain in knowledge. The three steps are repeated until no further hypotheses can be generated: The verified building hypotheses lead to predicted unobserved 2D primitives on the lower levels, giving additional information for reconstructing previously undetected corners, initiating a second iteration of 3D-reconstruction, generation of building hypotheses and verification.

3 Realization

3.1 Reconstruction of 3D Corners

In this section we describe the procedure for 3D-reconstruction and interpretation on the level of feature aggregates (cf. fig. 1). The reconstruction is based on local image aggregates A^{2D}, which can be directly derived from our polymorphic feature description. We distinguish corners and vertices, corners being interpreted vertices. As for vertex structures V^{2D}, being the projections of 3D vertices V^{3D} into the images, the imaging geometry gives strong restrictions during the correspondence analysis, we focus on the 3D reconstruction of corners C^{3D}.

Corner Model Corners are specialized into different subclasses. Each corner is described by the corner point, several lines and planar faces (cf. fig. 2). The partitioning of corners is described by their line attribution, given by the semantic labels (h), (v) or (o) (cf. Gülch 1992). This description is further refined by distinguishing vertical and oblique lines due to their slope into (v+), (v-), (o+) and (o-). Finally, different geometric constaints between the corner components like e. g. symmetry and orthogonality are considered.

We also model corner pairs. The previous classification of single corners is used for deriving restrictions like collinearity of lines and coplanarity of planes on compatible corner pairs.

All these semantic attributed subclasses are collected in the set Ω_C, one element for instance being the quadruple { (h), (o-), (o-), symmetry (o-,o-) }, which corresponds to the vertex at the ridge of a hip roof.

Construction of Corner Hypotheses Starting point for the analysis are the 2D vertices V^{2D} (cf. section 2.3). In the first step we generate 3D vertices. Selected tuples of corresponding vertex structures form the basis for the transition to 3D-vertices $V^{3D} = \{v_1^{3D}, \ldots, v_n^{3D}\}$ by a joint forward intersection of all corresponding 2D-vertices $\boldsymbol{v}^{2D} = (v_1^{2D}, \ldots, v_N^{2D})$. It is established by heuristically selecting a suitable sequence of vertices, which are then evaluated. The correspondence analysis uses the epipolar geometry and the structural similarity of matching candidates. Relational matching of the features F^{2D}, which describe the vertex structures v_i^{2D} of the correspondence tuple $\boldsymbol{v}^{2D}$, leads to a 3D Vertex v^{3D}.

Establishing corner hypotheses uses the corner model in the following way. We interpret the generated 3D vertices V^{3D} by a first *classification*, leading to admissible corner subclasses Ω_C. First, the line attributation is used to obtain a subset Ω'_C of Ω_C, which in case the line attributations do not correspond to a valid corner

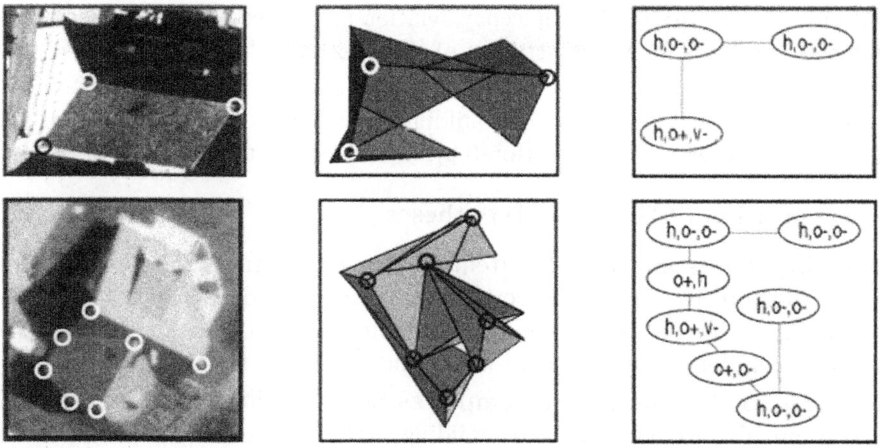

Fig. 7: Examples of 3D corner reconstruction. From left to right: feature aggregates marked in one image, reconstructed 3D corners, corner graph showing the adjacency relations.

description, is the empty set. For each element of Ω'_C the geometric constraints, associated by the model, yield a further reduced $\Omega''_C \subset \Omega'_C$, each element in Ω''_C being an admissable interpretation of the vertex v^{3D}. This way we obtain several corner hypotheses c^{3D}.

This principle of domain reduction of possible interpretations is also applied for the generation of corner pairs (c_i^{3D}, c_j^{3D}).

Verification of Corner Hypotheses As the corner classification for each vertex v^{3D} generally will be ambiguous, we perform a second rigorous classification by statistical analysis. This is an optimization problem for finding the best interpretation of the data. For each hypothesis, we estimate the geometric (and probably radiometric) parameters by a maximum likelihood parameter estimation. The functional model describes the relation between observations and the unknown parameters for the unconstrained corner. Additionally hard and soft constraints $\Theta^{3D}_{\omega_C}$, given by the corner class ω_C, are introduced. The result of the estimations are evaluated corner reconstructions $c^{3D} = c^{3D}(v^{3D}, \omega_C)$. Assuming the same image feature observations, all possible corners are tested against the unconstrained corner to decide for the optimal interpretation of the vertex data $\boldsymbol{v}^{2D}$.

Further details of the corner reconstruction approach can be found in (cf. Lang and Förstner 1996b, Brunn et al. 1996).

The geometric constraints on single corners c_i^{3D} as well as on corner pairs (c_j^{3D}, c_k^{3D}) are fundamental for a geometrically improved and verified corner reconstruction. Those corners C^{3D}, which are accepted, are regarded as 3D-observations of the level feature aggregates and form the basis for the 3D aggregation. The grouping of corners results in a weighted graph $G = (C^{3D}, R)$ where the nodes are classified corners and the arcs the adjacency relation. The weight of the arc expresses the conditional probability of the adjacency relation as derived from the classification. This forms the basis for the generation of building hypotheses as presented in the next chapter.

After the first iteration of the whole building reconstruction process, we use newly generated 2D corners C^{2D} (cf. section 3.3), in addition to the original vertex data.

3.2 Generation of Building Hypotheses

Building hypotheses are generated from the observed corner graph $G = (C^{3D}, R)$ in a two step process. First the corners are grouped and aggregated to building part templates. In the second step these building part templates are grouped and aggrated to building templates (cf. Fischer and Steinhage 1997).

Generation of building part templates Since building parts are defined by corner graphs G_j, the generation of building part templates explaining the corner observations is done by determination of maximal subsets $C_i^{3D\prime} \subseteq C^{3D}$ that have a subgraph isomorphism from $(C_i^{3D\prime}, R')$ onto a G_j. That is, corner observations and templates must have the same classifications and adjacency relations.

If a subgraph isomorphism of a $(C^{3D\prime}, R')$ in a graph G_j of a building part p is found, the elements of $C^{3D\prime}$ are aggregated to a new building part template $\dot{p}$.

Generation of building templates For the generation of building part templates structural and geometrical information are considered. In general not all building parts forming a building have corner observations. Therefore it may be necessary to hypothesize building part templates without having actual observations for them.

Buildings are constructed by grouping building parts. Our generic model allows groups of variable size as long as restrictions on pairs of building parts implied by the model are satisfied.

The grouping task is defined by iterating the following steps. Starting with the set T_0 of previously generated building part templates more complex templates are generated by merging compatible pairs of building parts. Merging two parts implies the unification of certain parameters and the introduction of a new length parameter (cf. fig. 8). Two building parts may be merged, if their plug faces are of the same type and they have compatible geometry. The latter is checked by parameter estimation. The previous step is iterated until either all elements of T_0 are part of a complete building with no plug faces or no new connections can be made.

If the first condition holds we are done, because the complete buildings in the final set T contain all elements of T_0 and therefore explain all corner observations. Otherwise,

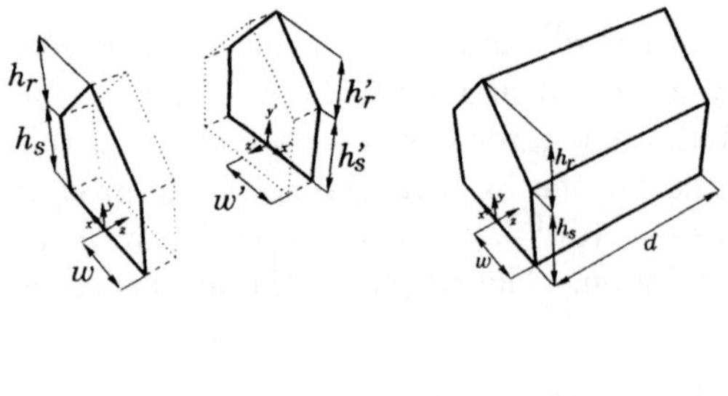

Fig. 8: The parameters w and w', h_s and h'_s, and h_r and h'_r are unified. The parameter d is introduced to describe the depth of the newly formed building.

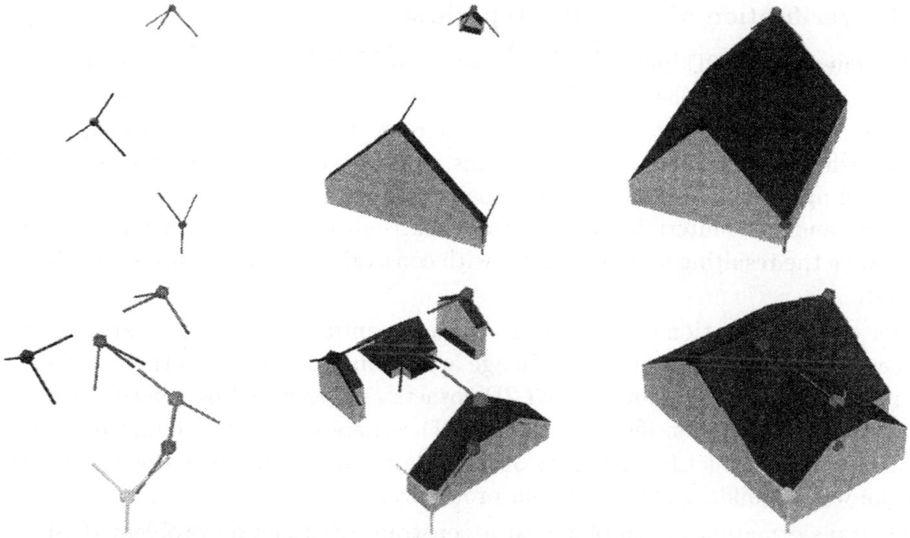

Fig. 9: Examples of building part aggregation. From left to right: reconstructed 3D corners, best matching building part hypotheses, aggregated building. Note that default values are used for free parameters.

not every building part template in T_0 is part of a complete building in T. In order to assign every element in T_0 to a complete building, new building part templates have to be introduced for which there exist no observations. These parts must enable the construction of new complete buildings, consisting of as few parts as possible. Every complete building represents a building template, which is a valid aggregation of building parts.

For every building template a global parameter estimation considering all associated corner observations determines the degree of freedom for every parameter.

Generation of 2D view hierarchy For verification and determination of free parameters of the generated building hypotheses, represented by the building templates, they have to be matched with the observations in the image. In 2D, buildings are modeled by view hierarchies. With respect to the fixed image geometry all topologically different aspects of the buildings correspond to primitives on the highest level of the view hierarchy, defining the 2D building templates. In contrast to aspect graphs the viewing direction is fixed, but due to free parameters the object geometry is in general variable, which can lead to different possible aspects. After the generation of the 2D building templates, they are decomposed into primitives and their interrelationships on lower levels. A relation between primitives on a higher level is transformed to a set of relations beween their components on the lower level. Thereby semantic knowledge is propagated down through every level of our aggregation hierarchy. For each image one view hierarchy is generated.

3.3 Verification of Building Hypotheses

The generation of 3D building hypotheses leads to different possible 2D views. Since some parameters of the 3D building templates may still be free, there are free parameters in the projected 2D views. Therefore the first task is to identify those views which match the given observations. A match fixes free parameters of the corresponding template. Since a match between template and observations considers objects and their interrelationships this task is an instance of relational matching. We solve the resulting search problem with constraint solving methods (cf. Meseguer 1989).

After the identification of the view, giving a semantic label to every matched feature observation, line fragments in the image are grouped according to the building template. Then previously unobserved 2D corners are generated by robust estimations on intersections of identified lines. These 2D corners are additional information that is used to reconstruct further, previously undetected 3D corners in the next iteration of the whole building reconstruction procedure.

The transformation of templates to a constraint satisfaction problem (CSP), the matching and its implementation using constraint logic programming (CLP) (cf. van Hentenryck 1989, Jaffar and Maher 1994) is explained in detail in Kolbe et al. 1996.

Transformation of view hierarchies to sets of constraints A view hierarchy enumerates the possible views for the building hypotheses with respect to one image. To evaluate the correspondence to the originally extracted image features the matching between the building templates and the observations is done on the lowest level of the hierarchy. The entities considered for matching are objects of the three feature classes points, lines, and regions and different relations, e. g. adjacency, line parallelism, region symmetry, and region contrast. The line parallelism and region symmetry relations reflect the respective 3D properties in 2D. The region contrast relation classifies the intensity ratio between adjacent regions.

A building template there is represented by a set of features and a set of relations between them. These relations may be regarded as a set of constraints which have to be satisfied simultaneously by the corresponding objects. A consistent match is an identification of a set of objects which satisfies all constraints. The decomposition of the matching problem into the simultaneous satisfaction of different constraints leads to a structure which is adequately represented by the translation of templates into conjunctions of constraints where the features are represented by variables with restricted domains. The domain of a variable depends on its associated model feature type, for instance line variables are restricted to the set of line observations that were extracted from the image. Thus the first constraint posed on a variable is the *domain constraint*. All other constraints are implied by the corresponding relations of the building template.

For each view of a hierarchy one set of constraints will be generated representing the building template.

Matching of constraint sets to the image data The matching is done seperately for every image and view by searching for a valid assignment of extracted

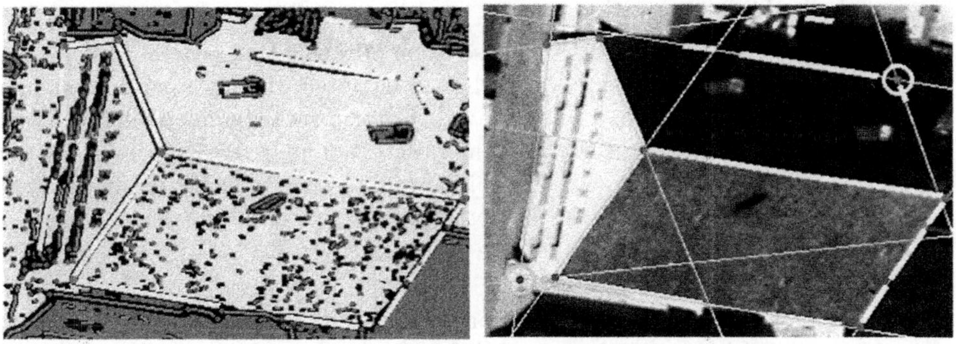

Fig. 10: Left: Matched features, right: new 2D corners

image features to the variables of the corresponding CSP, such that all constraints are satisfied. Consistency techniques exploit the structured representation of the matching problem as a constraint network. The idea is that a constraint is used to restrict the domains of its variables and that these reductions are propagated through the network.

Standard techniques for constraint solving (Mackworth 1977, Haralick and Elliott 1980, Dechter and Pearl 1988, van Hentenryck 1989) demand that a match satisifies every constraint. Due to occlusions and disturbances in general neither every predicted model feature of a view can be found nor all their incident constraints are satisfiable. To cope with this problem we employ relaxation of constraints, allowing constraints to be neglected. Our combination of consistency techniques and constraint relaxation follows in general the scheme described in Haralick and Shapiro 1993 for inexact matching. As a language constraint logic programming provides hard constraints which are strictly enforced. A soft constraint $c(t_1, \ldots, t_{n-1})$ with $n-1$ arguments may be implemented by a hard constraint $c'(t_1, \ldots, t_{n-1}, b)$ where c' is satisfied if the boolean b is $true$ or c is satisfied. The variable b with the domain $\{true, false\}$ may act as a communication channel between the constraint and its environment. This environment may come to the conclusion that c models an unobservable relation and relaxes c by instantiating b with $true$.

The solution of the constraint satisfaction problem are the valid assignments of extracted features to the variables. On the one hand these assignments determine the free geometric parameters of the feature templates and thus determine free parameters of the whole building template. On the other hand, every assigned image feature is identified as a specific component of the building observation as defined by the building template. These matched image features are given labels, denoting their semantics with respect to the building model. Figure 10 shows on the left highlighted the image features that were successfully matched with the building template for a saddle roof house.

Since lines in general are fragmented into several line image features, corresponding line features are grouped and fitted using the labeling information to determine the lines of the building observation.

Generation of new 2D corners The solution of the CSP in general fixes further parameters of the building template. It has to be noted, that free parameters may remain, if some constraints had to be relaxed. Corners of the building template with fixed positions that have no corresponding observation of a vertex structure in the image, finally are reconstructed in 2D as illustrated by the right picture in figure 10 to provide new information for the 3D corner reconstruction in the next iteration of the reconstruction process.

As mentioned above the loop of 3D-reconstruction and generation and verification of building hypotheses will be terminated if no further hypotheses can be generated. After the final verification step the parameters of the 3D-building model will be determined simultaneously based on the observed image features and taking all geometric and possibly radiometric constraints into account.

3.4 Results

The actual result of the reconstruction procedure is presented for the international test data set which was distributed for the Ascona Workshop 1995 on *Automatic Extraction of Man-Made Objects from Aerial and Space Images* (cf. Grün et al. 1995). The image scale is 1 : 5000 and we use a resolution with $30\mu m$ pixel size in contrast to Bignone et al. 1996 who use $15\mu m$. As building **h12** is under construction, only 11 out of the 12 buildings which are contained in multiple images, are relevant for the analysis. To test the feasibility of the concept in a first instance the number of building primitives is reduced to 2 different building part terminals and their possible connectors. To resolve the footprint of the buildings, higher image resolution is necessary. Thus building heights can not be determined automatically and therefore are set to fixed values. Figure 11 shows the result after the first iteration loop (cf. Fig. 6). Results of the intermediate steps are listed in Table 1 and reflect the following aspects of the different reasoning steps:

Reconstruction of 3D corners (I): The number of reconstructed corners `RC` on average compounds 40% of the buildings corners (`C`), whereas 2 of them are incorrectely classified. We are not able to completely reconstruct every corner, because not all corners are observable in the symbolic image descriptions. An application indepent grouping as presented in Fuchs and Förstner 1995 may improve the initial symbolic image descriptions. Possibly still remaining unreconstructed corners have to be identified during the verification step of building hypotheses. The reason for incorrect classification is given by a weak intersection geometry for the line reconstruction, which is not yet considered in the modeling process. Neglecting the influence of the uncertainity of the line reconstruction, the corner point is correctely reconstructed. Actually the identification of wrong corner interpretations is performed by finding global inconsistencies during the generation of building hypotheses. For building `h10` the image resolution is not sufficient to successfully reconstruct any corner. The parameter estimation for correctely classified corners

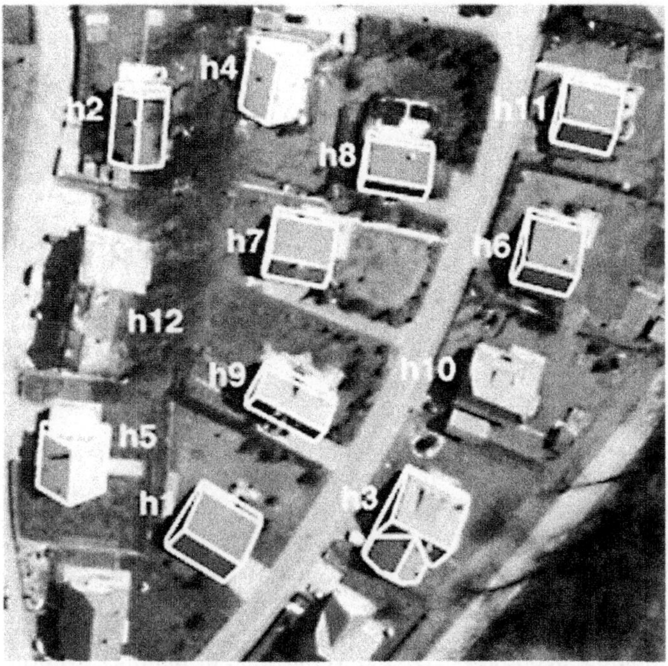

Fig. 11: Visualization of the reconstruction result for the Ascona Dataset after one iteration loop (cf. Fig. 6) is performed. The images contain 12 buildings in multiple overlap. As building h12 is under construction, we excluded it from our analysis. We are able to completely reconstruct 10 of the 11 buildings which were relevant for the analysis.

using 4 images simultaneously achieves an accuracy of the reconstructed corner point about $\sigma_x = \sigma_y = \pm 6cm$ and $\sigma_z = \pm 15cm$. The accuracy of the orientation of the corners is about $\sigma_\kappa = 0.6^o$ while the reconstruction of the slope of the roof is by $\sigma_{sl} = 2^o$.

Generation of building hypotheses (II): The column BPH gives the number of building part hypotheses which are generated for each of the reconstructed single corners. The combination of these building parts results in the number BH of hypotheses of complete buildings which are consistent with all reconstructed corners. If only one corner is given, as it is the case for h2 and h7, the corresponding terminals are each completed to a closed building hypothesis with a second terminal of the same type. For each of the BH building hypotheses the view graph is generated. Currently the view graph of the best building hypothesis is passed to the verification of building hypotheses. The decision which is the best one is given by using the Minimum Description Length criterium following Rissanen 1987.

Verification of building hypotheses(III): For each building in each image one view graph being a building template is available and matched to the extracted image features and mutual relations. Column UV lists the number of corners which were correctly predicted during the generation of building hypotheses. The column GV shows for each of the 4 images the number of correctly matched vertices. For 80% of the images a successful matching is reached, 9 of the 10 generated building hypotheses could be verified by successful matching in at least one image. Dashes in the table denote a break-off of the analysis with complex or strongly undercon-

	I			II		III		
Building	C	RC	CE	BPH	BH	UV	GV	FP
h1	6	3	0	3/4/3	4	3	3/3/3/3	0
h2	6	1	0	3	2	5	5/5/-/-	0
h3	12	7	0	3/3/2/3/3/4/0	4	5	-/-/-/-	0
h4	6	3	0	4/3/3	4	3	3/3/3/3	0
h5	6	3	0	3/4/4	4	3	3/3/3/3	0
h6	6	3	1	3/0/4	4	4	4/4/4/4	0
h7	6	2	1	3/0	2	5	5/5/5/5	0
h8	6	3	0	3/3/4	4	3	3/3/3/3	0
h9	6	4	0	3/3/4/8	2	2	2/2/2/2	0
h10	10	0	-	-	-	-	-	-
h11	6	2	0	3/3	4	4	4/4/1/0	0

C number of corners of the building
RC number of reconstructed corners
CE number of corner classification errors
BPH number of building part hypotheses for each corner
BH number of building hypotheses
UV predicted corners
GV predicted and verified vertices per image
FP remaining free parameters

Tab. 1: Detailed results of the intermediate reconstruction processes Reconstruction of 3D corners (I), Generation of building hypotheses (II) and Verification of building hypotheses(III) for the Ascona Dataset. 10 of the 11 buildings can be reconstructed as shown in column FP.

strained models, which can be resolved using appriopriate heuristics to keep the search space down. As for the buildings h3 the reconstructed corners already were sufficient to determine all parameters of the building hypothesis, in 10 of the 11 examples the building is successfully reconstructed as shown in the last column FP of the table.

4 Conclusions and Outlook

We have proposed a model-based approach to 3D building extraction from aerial images. Our modeling concept reveals a tight coupling of generic 3D object modeling and an explicit image model. Object and image model show corresponding *part-of* hierarchies describing aggregation states within a recognition-by-components strategy.

The procedure aims at a step by step increase of knowledge during the reasoning. As the semantics of a model defines the maximum achievable knowledge about an actual scene and its granularity determines the resolution for knowledge our model spans hierarchically from the smallest observable features to complex buildings and allows an adequate evaluation of hypotheses and their immanent knowledge within the different reasoning steps.

We have presented first experimental results on test data sets provided by the Lan-

desvermessungsamt Bonn and the ETH Zürich.
Currently we are investigating and developing on

- the measurement and propagation of uncertainty within the overall reconstruction process;
- the extension of the current building modeling by more sophisticated knowledge about buildings, esp. functional aspects.
- the derivation of domain dependent heuristics to constrain search spaces;
- the handling of incomplete observations within all processes and stages of hypotheses generation and verification.

Especially we have to consider the integration of hard and soft constraints, i. e. logical and statistical knowledge into the framework of constraint logic programming. Furthermore the approach has to be compared to different automatic algorithms for segmentation, 3D reconstruction and classical photogrammetry. In order to achieve this goal an attempt is made to find a set of standardized notions about basic building models, parts of buildings and classes of buildings to ease such comparisons on a conceptual level.

Acknowledgments:

This work was largely done within the the project "Semantic Modeling and Extraction of Spatial Objects from Images and Maps" especially in the subproject "Building Extraction" which is supported by the Deutsche Forschungsgemeinschaft. We heavily profitted from discussions with Armin B. Cremers, Wolfgang Förstner, Lutz Plümer and Volker Steinhage. We thank the DFG for supporting our work.

References

Bergevin R., M. Levine (1993) *Generic object recognition: Building and matching coarse descriptions from line drawings*, IEEE T-PAMI, Vol. 15, No. 1, pp. 19–36.

Biederman I. (1987) *Recognition-by-components: A theory of human image understanding*, Psychological Review, Vol. 94, pp. 115–147.

Bignone F., O. Henricsson, P. Fua, M. Stricker (1996) *Automatic extraction of generic house roofs from high resolution aerial images*, Computer Vision — ECCV '96, Proc. of the 4th Europ. Conf. on Computer Vision, number 1064 in *Lecture Notes in Computer Science*, pp. 85–96.

Braun C. (1994) *Interpretation von Einzelbildern zur Gebäudeerfassung*, PhD thesis, Institut für Photogrammetrie, Universität Bonn.

Braun C., T. H. Kolbe, F. Lang, W. Schickler, V. Steinhage, A. B. Cremers, W. Förstner, L. Plümer (1995) *Models for photogrammetric building reconstruction*, Computer & Graphics.

Brunn A., F. Lang, W. Förstner (1996) *A procedure for segmenting surfaces by symbolic and iconic image fusion*, in B. Jähne, P. Geißler, H. Haußecker, F. Hering (eds), Mustererkennung 1996, DAGM, Springer, pp. 11–20.

Clowes M. (1971) *On seeing things*, Artificial Intelligence, Vol. 2, pp. 79–116.

Dechter R., J. Pearl (1988) *Network-based heuristics for constraint satisfaction problems*, Artificial Intelligence.

Dickinson S. J., A. P. Pentland, A. Rosenfeld (1992) *From volumes to views: An approach to 3-d object recognition*, CVGIP: Image Understanding, Vol. 55, No. 2, pp. 130–154.

Fischer A., V. Steinhage (1997) *Solid modeling for building extraction from aerial images*, in V. Skala (ed.), 5th Int. Conf. in Central Europe on Computer Graphics and Visualization 97, WSCG'97, pp. 114–123.

Flynn P. J., A. K. Jain (1991) *Cad-based computer vision: From cad-models to relational graphs*, IEEE T-PAMI, Vol. 13, No. 2, pp. 114–132.

Förstner W. (1994) *A framework for low level feature extraction*, in J.-O. Eklundh (ed.), Computer Vision, ECCV '94, Vol. II, number 801 in *Lecture Notes in Computer Science*, Springer-Verlag, pp. 383–394.

Fua P., A. J. Hanson (1987) *Resegmentation using generic shape: Locating general cultural objects*, Pattern Recognition Letters, Vol. 5, pp. 243–252.

Fuchs C., W. Förstner (1995) *Polymorphic grouping for image segmentation*, Proceedings 5th International Conference on Computer Vision, pp. 175–182.

Grün A., O. Kübler, P. Agouris (eds) (1995) *Automatic Extraction of Man-Made Objects from Aerial and Space Images*, Birkhäuser.

Gülch E. (1992) *A knowledge-based approach to reconstruct buildings in digital aerial imagery*, 17th ISPRS Congress, Washington D. C., Proceedings, pp. 410–417.

Hansen C., T. Č. Henderson (1993) *Cad-based computer vision*, IEEE T-PAMI, Vol. 11, pp. 1187–1193.

Haralick R. M., G. L. Elliott (1980) *Increasing tree search efficiency for constraint satisfaction problems*, Artificial Intelligence, Vol. 14, pp. 263–313.

Haralick R. M., L. G. Shapiro (1993) *Computer and Robot Vision*, Vol. II, Addison-Wesley Publishing Company.

Herman M., T. Kanade (1987) *The 3d mosaic scene understanding system: Incremental recognition of 3d scenes from complex images*, in Fischler/Firschein (ed.), Readings in Computer Vision, Kaufmann, pp. 471–482.

Heyden A. (1994) *Consistency and correction of line-drawings, obtained by projections of piecewise planar objects*, Computer Vision — ECCV '94, Proc. of the 3rd Europ. Conf. on Computer Vision, number 800 in *Lecture Notes in Computer Science*, pp. 411–419.

Huertas A., M. Benjanin, R. Nevatia (1995) *Model registration and validation*, in A. Grün, O. Kübler, P. Agouris (eds), Automatic Extraction of Man-Made Objects from Aerial and Space Images, Birkhäuser, Basel, pp. 33–43.

Huffman D. A. (1971) *Impossible objects as nonsense sentence*, Machine Intelligence 6, pp. 295–323.

Ikeuchi K., P. J. Flynn (1995) *Recent progress in cad-based vision*, CVGIP, Vol. 61, No. 3, pp. 293–294.

Jaffar J., M. J. Maher (1994) *Constraint logic programming: A survey*, Journal of Logic Programming.

Kanatani K. (1990) *Group-Theoretical Methods in Image Understanding*, Springer.

Kolbe T. H., L. Plümer, A. B. Cremers (1996) *Using constraints for the identification of buildings in aerial images*, Proceedings of the 2. Int. Conf. on Practical Applications of Constraint Technology PACT'96 in London, The Practical Application Company Ltd., pp. 143–154.

Lang F., W. Förstner (1996a) *3d-city modelling with a digital one-eye-stereo system*, ISPRS Symposium, Comm. IV, Vienna.

Lang F., W. Förstner (1996b) *Surface reconstruction of man-made objects using polymorphic mid-level features and generic scene knowledge*, Zeitschrift für Photogrammetrie und Fernerkundung, Vol. 6, pp. 193–201.

Lin C., A. Huertas, R. Nevatia (1994a) *Detection of buildings using perceptual grouping and shadowsproceedings computer vision and pattern recognition*, Proceedings Computer Vision and Pattern Recognition, pp. 62–69.

Lin C., A. Huertas, R. Nevatia (1995) *Detection of buildings from monocular images*, Automatic Extraction of Man-Made Objects from Aerial and Space Images, pp. 125–134.

Lin C. L., Q. Zheng, R. Chelappa, L. S. Davis, X. Zhang (1994b) *Site model supported monitoring of arial images*, Proceedings Computer Vision and Pattern Recognition, pp. 694–699.

Mackworth A. K. (1977) *Consistency in networks of relations*, Artificial Intelligence, Vol. 8, pp. 99–118.

McKeown D. M. (1990) *Towards automatic cartographic feature extraction from arial imagery*, in H. Ebner, D. Fritsch, C. Heipcke (eds), Digital Photogrammetric Systems, Wichmann, Karlsruhe.

Meseguer P. (1989) *Constraint satisfaction problems: An overview*, AICOM, Vol. 2, No. 1, pp. 3–17.

Munkelt O. (1995) *Aspect-trees: Generation and interpretation*, CVGIP, Vol. 61, No. 3, pp. 365–386.

Quam L., T. Strat (1991) *Sri image understanding research in cartographic feature extraction*, in C. H. H. Ebner, D. Fritsch (ed.), Digital Photogrammetric Systems, Wichmann, Karlsruhe, pp. 111–121.

Rissanen I. (1987) *Minimum description length principle*, Encyclopedia of Statistical Sciences, Vol. 5, pp. 523–527.

Schickler W. (1993) *Towards automation in photogrammetry*, Geodetical Info Magazine.

Sester M., W. Förstner (1989) *Object location based on uncertain models*, in H. Burkhardt, K.-H. Höhne, B. Neumann (eds), Mustererkennung 1989, 11. DAGM-Symposium, Hamburg, Vol. 219 of Informatik Fachbericht, Springer, pp. 457–464.

Suetens P., P. Fua, A. J. Hanson (1992) *Computational strategies for object recognition*, ACM Computing Surveys, Vol. 24, No. 1, pp. 5–61.

Sugihara K. (1986) *Machine Interpretation of Line Drawings*, MIT-Press.

van Hentenryck P. (1989) *Constraint Satisfaction in Logic Programming*, Logic Programming Series, MIT Press, Cambridge, MA.

Waltz D. L. (1975) *Understanding line drawings of scenes with shadows*, in P. H. Winston (ed.), Psychology of Computer Vision, McGraw-Hill, New York, pp. 19–91.

Weidner U. (1996) *An approach to building extraction from digital surface models*, ISPRS.

Weidner U., W. Förstner (1995) *Towards automatic building extraction from high resolution digital elevation models*, ISPRS Journal, Vol. 50, No. 4, pp. 38–49.

Semanitic Modeling for the Acquisition of Topografic
Information from Images and Maps
ed. by W. Förstner and L. Plümer

Systematic Acquisition of Generic 3D Building Model Knowledge

Roman Englert

Rheinische Friedrich-Wilhelms-Universität Bonn
Institut für Informatik III
Römerstraße 164, D-53117 Bonn, Germany

englert@cs.bonn.edu

Abstract

3D modeling of man-made objects depends strongly on models. In this paper we present an approach for systematic acquisition of model knowledge from a given sample set using parameterized graphs and graph p-subisomorphism computation. Since subgraph isomorphism computation is well-known to be NP-hard background knowledge is used in order to guide the search for subgraphs and furthermore two pruning mechanisms are described. The approach will be evaluated by two different tasks using a large set of 3D building descriptions: (a) most frequent occuring parts of roofs, and (b) 3D building corners which are computed and grouped according to their geometrical and topological characteristics.

Keywords: Model Knowledge Acquisition, Automatic Extraction of Man-Made Objects, Subgraph Isomorphism, Symbolic Scene Description

1 Introduction

Three-dimensional modeling of man-made objects is an important endeavor which has to be solved for environmental construction and planning (Faugeras 1993). It is more challenging for objects modeled from images since problems as occlusions, inverse mapping, and noise (Braun et al. 1995) have to be tackled. A variety of approaches exist for the automatic extraction of man-made objects from images (Grün et al. 1995). Most approaches require model knowledge and thus their capability of object extraction depends strongly on the quality of the model knowledge used, or with other words: using sparse model knowledge which does not contain all features of the objects to be modeled would result in incomplete modeling (Besl and Jain 1985).

In this paper we present an approach for systematic acquisition of model knowledge based on a large given data sample by using background knowledge. We will illustrate our approach using 3D building surface descriptions wherever an increased demand on 3D building data has been observed (Fuchs 1996).

The acquisition of building model knowledge (short: building models) has multiple requirements: building models must include parts of roofs (Brunn et al. 1997), topological- and geometrical information must be covered, isomorphic structures must be computable, the acquisition must be repeatable and retraceable, etc.

Most man-made objects, esp. complex buildings, can be modeled by combinations of polyhedra (Englert and Gülch 1996). Unfortunately, they lack geometrical and topological information, e.g. symmetry, size of faces or orthogonality of neighbored faces. Therefore, as an appropriate representation for 3D buildings *parameterized graphs* will be defined (Section 2) which enable one explicitly to model additional knowledge. From parameterized graphs maximal subgraph isomorphisms can be computed systematically using parameter information added to the graphs and background knowledge, where 'maximal' refers to the size of the computed subgraphs (Section 3). Maximal subgraph isomorphisms will serve as candidates for generic 3D building model knowledge. For their computation background knowledge will be used which has to be defined by an user. Subgraph isomorphism testing is well-known to be NP-complete. The use of background knowledge is recommended in order to guide the search, and thus to reduce computational complexity. The computed maximal subgraph isomorphisms represent building models. In Section 4 the systematic extraction of building models and its practical use is demonstrated by two different tasks: (a) building models which are maximal subgraphs are computed and grouped into different classes according to their size, and (b) 3D corners of buildings are extracted and grouped according to their topological characteristics. Both tasks are performed on a large sample set consisting of 1846 groups of 3D buildings which represent the suburban area Oedekoven close to Bonn. Finally, we conclude with a discussion of the future work.

2 Basic Notion of Parameterized Graphs

In this section *parameterized graphs* and *p-subisomorphic* graphs will be defined. Basic concepts of graph theory can be found in, e.g. (Aigner 1988, Harary 1972).

1 DEFINITION (PARAMETER VECTOR)
A PARAMETER VECTOR $A = (a_1, \ldots, a_n)$ *of length n is a vector consisting of n real-valued constants or real-valued variables.* a_i *is called* COMPONENT OF A.

2 DEFINITION (PARAMETERIZED GRAPH, P-GRAPH)
A PARAMETERIZED GRAPH[1] $G_P = (V, \zeta, E, \xi)$ *(short:* P-GRAPH*) is an undirected graph, where each node* $v \in V$ *has a parameter vector* $\zeta(v)$ *and each edge* $e \in E$ *has a parameter vector* $\xi(e)$. *The* NEIGHBORHOOD $N(V')$ *of a node set* $V' \subseteq V$ *is* $N(V') \stackrel{def.}{=} \{v \in V \backslash V' | \; \exists w \in V' : \; \{v, w\} \in E\}$ *and the* DEGREE $\delta(V')$ *is* $|N(V')|$.

[1]The concept of parameterized graphs is borrowed from (Mehlhorn and Näher 1995). In contrast, we allow the occurrence of variables in parameter vectors instead of constants.

Note, all parameter vectors of a p-graph can be of different length. On the other hand, parameter vectors do not permit the discrimination of nodes, resp. edges, from each other. The following extensions of graph concepts are based on *parameter vectors*. The term $|\cdot|$ denotes the cardinality of a set:

3 DEFINITION (p_n-CONNECTED)
*A p-graph $G_P = (V, \zeta, E, \xi)$ is called p_n-*CONNECTED *iff. for all pairs of nodes $\{v, w\} \in V \times V$ exists a path $P \in E^k, k \leq |E|$ from v to w, that each edge $e \in E$ has a parameter vector $A = \xi(e)$ of length n. Two edges $e_1 = \{v, w\}$ and $e_2 = \{p, q\}$ are* ADJACENT *iff. $v = p$ and $w \neq q$.*

Since nodes of a p-graph have a parameter vector, we assume that p-graphs do not entail self-loops. Analogously, p-graphs must not have parallel edges. The notion of an *induced p-subgraph* enables one to extract a subgraph from a p-graph maintaining its information given by parameter vectors.

4 DEFINITION (IS SUBSUMED BY, $\prec$)
Given two parameter vectors $A = (a_1, \ldots, a_n)$ and $B = (b_1, \ldots, b_n)$ of length n. A IS SUBSUMED BY *B, written $A \prec B$, iff. for $i = 1, \ldots, n$ holds $a_i = b_i$ or a_i is a variable and b_i a constant.*

5 DEFINITION (IS EQUIVALENT TO, $\doteq$)
Two parameter vectors A and B are called EQUIVALENT, *written $A \doteq B$, iff. $A \prec B$ and $A \succ B$.*

The trivial case in which each parameter vector is subsumed by itself will not be considered. As example take the parameter vectors $A = (a, X)$ [2], $B = (X, b)$, $C = (Y, X)$, and $D = (a, c)$, then $A \preceq D$, $C \preceq A$, $C \preceq B$, and $C \prec D$ hold, where $P \preceq P'$ means $P \prec P'$ or $P \doteq P'$.

Subgraphs of p-graphs have to take relations of parameter vectors of nodes and edges into consideration. Note, the *binomial coefficient* $\binom{S}{k}$ denotes the sets consisting of k objects without repetition out of a set S, and the mapping $f|M$ denotes the *restriction of f to M*, where $f|M : M \to Y$, with $M \subset X$, assuming a mapping $f : X \to Y$ is given.

6 DEFINITION (P-SUBGRAPH)
A P-SUBGRAPH *$G'_P = (V', \zeta', E', \xi')$ of a parameterized graph $G_P = (V, \zeta, E, \xi)$ is a p_n-connected p-graph with $V' \subseteq V$, $E' \subseteq E \cap \binom{V'}{2}$, $\zeta' = \zeta|V'$, and $\xi' = \xi|E'$. G'_P is called* INDUCED P-SUBGRAPH OF *G_P, iff. $E' = \binom{V'}{2} \cap E$.*

[2] Lower letters denote constants and upper letters denote variables.

7 DEFINITION (UNION, $\dot{\cup}$)
Given two induced p-subgraphs $G'_P = (V', \zeta', E', \xi')$ and $G''_P = (V'', \zeta'', E'', \xi'')$ of a p-graph $G_P = (V, \zeta, E, \xi)$, with $V' \cap V'' = \emptyset$. Then G'_P UNION *G''_P, written $G'_P \dot{\cup} G''_P$, denotes the p-graph $G_{\dot{\cup}} = (V_{\dot{\cup}}, \zeta_{\dot{\cup}}, E_{\dot{\cup}}, \xi_{\dot{\cup}})$, with $V_{\dot{\cup}} = \{V' \dot{\cup} V''\}$, $\zeta_{\dot{\cup}} = \zeta | V_{\dot{\cup}}$, $E_{\dot{\cup}} = \{\binom{V' \dot{\cup} V''}{2} \cap E\}$, and $\xi_{\dot{\cup}} = \xi | E_{\dot{\cup}}$.*

8 DEFINITION (P-ISOMORPHIC, $\simeq_P$)
A p-graph $G_P = (V, \zeta, E, \xi)$ is P-ISOMORPHIC *to a p-graph $G'_P = (V', \zeta', E', \xi')$, written $G_P \simeq_P G'_P$, iff. there exist bijections $\phi_V : V \to V'$ and $\phi_E : E \to E'$, with $\phi_E(\{v, w\}) = \{\phi_V(v), \phi_V(w)\}$ for all edges $\{v, w\} \in E$, with $\zeta(v) \preceq \zeta'(\phi_V(v))$ for all nodes $v \in V$, and with $\xi(e) \doteq \xi'(\phi_E(e))$ for all edges $e \in E$.*

As a consequence, a node v is p-isomorphic to a node v', written $v \simeq_P v'$, if for their embedding p-graphs $G_P = (\{v\}, \zeta|\{v\}, \emptyset, \varepsilon_f)$ [3] and $G'_P = (\{v'\}, \zeta'|\{v'\}, \emptyset, \varepsilon_f)$ holds $G_P \simeq_P G'_P$. For practical reasons we will abbreviate $G_P \simeq_P G'_P$ by the quadruple $[G_P, G'_P, \phi_V, \phi_E]$ and $v \simeq_P v'$ by the quadruple $[v, v', \phi_V, \phi_E]$. Analogously, $G_P \dot{\cup} v'$ denotes G_P UNION G'_P. The concepts of an *admissible extension* of induced p-subgraphs and *p-subisomorphic* graphs are at the core of the approach presented.

9 DEFINITION (ADMISSIBLE EXTENSION)
Given two p-graphs $G_P = (V, \zeta, E, \xi)$ and $G'_P = (V', \zeta', E', \xi')$ and two nodes $v \in V$ and $v' \in V'$. The node pair (v, v') is an ADMISSIBLE EXTENSION *of the pair (G_P, G'_P), iff. $v \notin V$, $v' \notin V'$, and $G_P \dot{\cup} v$ is p-isomorphic to $G'_P \dot{\cup} v'$.*

0 DEFINITION (P-SUBISOMORPHIC, $\sim_P$)
A p-graph $G_P = (V, \zeta, E, \xi)$ is P-SUBISOMORPHIC *to a p-graph $G'_P = (V', \zeta', E', \xi')$, written $G_P \sim_P G'_P$, iff. there exists an induced p-subgraph G^i_P of G'_P which is p-isomorphic to G_P, where the upper index i denotes the number of nodes of G^i_P. G_P is called* TEMPLATE GRAPH *and G'_P* SEARCH GRAPH.

3 Computation of Maximal Subgraph Isomorphisms

In the previous section notions are defined in order to represent man-made objects and background knowledge including geometric and topological information. The task to solve is the following: given two p-graphs S and T, find maximal p-subgraphs T' of T which are induced p-subgraphs of S, where 'maximal' refers to the number of nodes of T. Thus all T' are p-subisomorphic to S and T. As an example consider Figure 1: in case (a) T'_1 is p-isomorphic to T_1. The parameter vector $(W4, 4)$ of S_1 prevents the existence of a second maximal induced p-subgraph of S_1. Therefore in case (b) the parameter vector $(4, V5)$ of S_2 prevents the existence of a p-isomorphic p-graph of S_2. But there exist two different induced p-subgraphs T'_2 of S_2 both of which having two nodes.

[3] ε_f denotes the so-called *empty function* $f : \emptyset \to \emptyset$

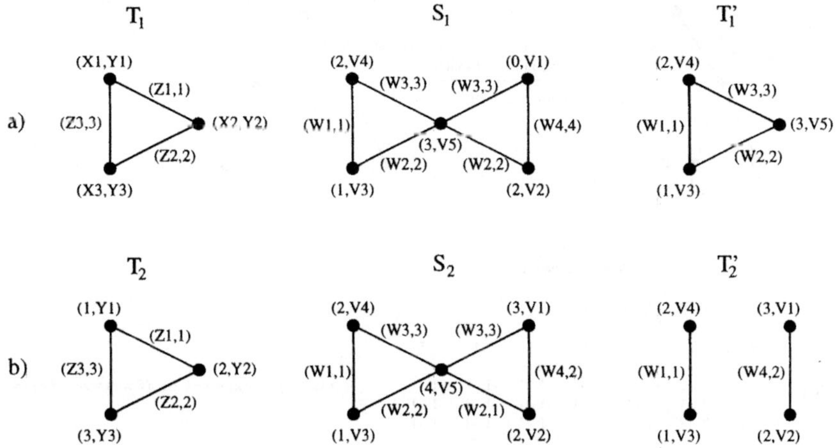

Fig. 1: The p-graphs T_1' and T_2' are p-subisomorphic to S_1 or T_1 and S_2 or T_2. In example (a) T_1' is p-isomorphic T_1; the parameter vector $(W4, 4)$ of S_1 prevents the existence of a second maximal induced p-subgraph of S_1 and in example (b) the parameter vector $(4, V5)$ of S_2 prevents the existence of a p-isomorphic graph of T_2; but two different induced p-subgraphs T_2' of S_2 exist both of which having two nodes.

The subgraph isomorphism problem is well-known to be NP-complete. In the case of planar graphs linear algorithms for isomorphism computation are known (Fontet 1976, Hopcroft and Wong 1974, Ullmann 1976). These approaches have two demerits: (a) during computation they do not allow to perform constraint checking in order to incorporate geometrical and topological information, and (b) the search of subgraph isomorphisms has to be performed top-down according to the number of nodes by testing all subsets with n nodes, $n-1$ nodes, etc. resulting in exponential computation. The latter case requires tremendous efforts with increasing difference of the number of nodes of both graphs.

We propose an algorithm which performs a bottom-up computation and additionally prunes the search space by eliminating permutations as early as possible of their occurrence and performing a look ahead. The main idea is to perform a breadth-first search traversing the p-graph along all incident edges given a start node: take a node n_T of T as start node, match n_T to all possible nodes n_S of S according to their parameter vectors, and try to match each neighbor of n_T to all neighbors of n_S according to their parameter vectors, etc.

Assume T has N nodes. Each matching will be stored in a dynamic $K \times N$ matrix I (cf. Figure 2), where K denotes the rows of I and N the columns of I. Each row of I contains exactly one computed p-subgraph, and each column consists of two indices of the nodes n_T and its matching counterpart n_S. The algorithm takes as input two p_n-connected p-graphs T and S and computes the maximal induced p-subgraphs T' of T which are p-subisomorphic to S. First, we will describe the

(a)

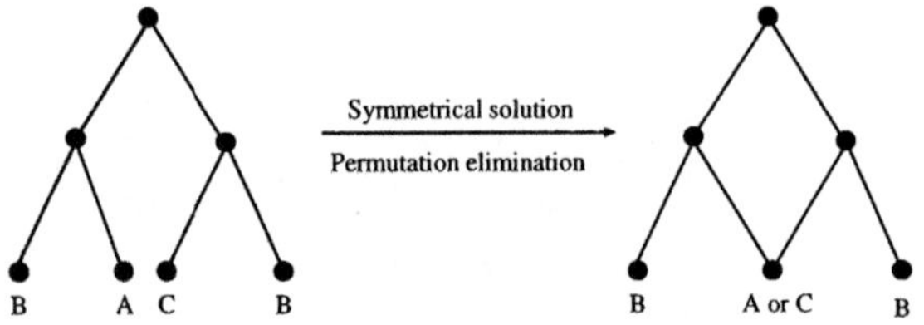

Matching matrix

(b)

	T' S	T' S	T' S	...			
Index	0	1	2	3	4	5	
0	0 0	-1 -1	-1 -1	-1 -1	-1 -1	-1 -1	matching of start node 0 fromT with node 0 from S
1	0 1	-1 -1	-1 -1	-1 -1	-1 -1	-1 -1	
2	0 2	-1 -1	-1 -1	-1 -1	-1 -1	-1 -1	
...							
5	0 5	-1 -1	-1 -1	-1 -1	-1 -1	-1 -1	matching of start node 0 fromT with node 5 from S
6	0 0	2 2	3 3	1 1	5 5	4 4	matching of neighbor nodes of node 0 from T
7	0 0	2 3	3 2	1 1	5 5	4 4	and a permutation of nodes 2 and 3 or of all
8	0 1	2 3	3 2	1 5	5 0	4 4	nodes except node 4
...							

Fig. 2: Graphical- and numerical examples of permutations. In part (a) we assume that the left tree contains symmetrical solutions A and C, where C is a permutation of A and thus can be eliminated by merging both paths as indicated in the right tree. In part (b) the head of a matching matrix I is shown. Row 7 is a permutation of row 6 and thus row 7 can be eliminated as soon as column 3 has been computed.

algorithm without pruning mechanisms:

1. Initialize the variables matching matrix $I = 0$, level of matching $ML = 0$, last column of I $LCI = 0$, last row of I $LRI = 0$.

2. Assign LRI to IA. Take n_T from T as start node and try to match it with all nodes n_S of S, protocol each successful matching in I by appending the indices of n_T and n_S to row LRI and increment LRI after each successful matching.

3. If $LRI = 0$ then no matching with node index n_T can be done: go to step 5. Otherwise increment LCI.

4. Repeat N times: increment ML and traverse I in a vertical manner from row IA to row LRI computing the following matchings for each row: for all neighbors nn_T of the matched nodes n_T from T in the previous pass, try to match nn_T with all neighbors nn_S of the node index n_S from S which were matched with n_T in the previous pass and store each successful matching in I by incrementing LRI, copying the current row to row LRI and appending the pair n_T and n_S to row LRI.

5. If a node n_T from T has not yet been used as start node, then go to step 2 and use n_T as start node.

6. Check if all edges have been matched.

7. Eliminate permutations of p-isomorphic graphs.

8. Sort rows of I by number of used columns and print I.

The main work of the algorithm is done in step 4: the protocol I is traversed vertically by testing if the p-subgraphs stored in the rows can be expanded by further matchings with nodes from S. Therefore, each row will be processed in a horizontal manner trying to match the nodes matched during the last pass. Thus the algorithm computes all possible induced p-subgraphs T' of S by starting with a node n_T and appending all successfully matched nodes nn_T with corresponding nodes nn_S of S. Note, a matching is only successful if the parameter vector $\xi(nn_T)$ is subsumed by/or equal to $\xi(nn_S)$. Only node indices are stored. Since the p-graphs are traversed along adjacent edges a connected p-subgraph will be computed. Finally, the edges of the matched nodes have to be checked, since the edges of neighbored nodes which were matched in earlier passes are not checked during p-subgraph isomorphism computation. Afterwards, permutations of node indices of p-isomorphic graphs will be eliminated and I will be printed in a sorted order.

Each successful node matching is stored in I by copying the current processed row to a new row appended to I, and appending the node indices of the matched nodes after the last column of the new row. Having processed step 4 two effective prunings of I, which also prune the search space, can be performed.

Pruning mechanism SYMMETRICAL SOLUTION: Rows created or entailed during the last pass can be eliminated if they are a permutation of any other row of I which has at least the same number of used columns. As example consider Figure 2: a graphical example is shown in part (a). We assume that the left tree contains symmetrical solutions A and C, where C is a permutation of A and thus can be eliminated by merging both paths as indicated in the right tree. In part (b) the head of a matching matrix I is shown: row 6 is a permutation of row 7 and thus row 7 can be eliminated as soon as column 3 has been computed. Note, this pruning mechanism does not hold for directed graphs.

Pruning mechanism LOOK AHEAD: If a node has several neighbor nodes which are not already matched at the current computation state, then it suffers to protocol their successful matching in the current row without copying it. As example consider the matching matrix (part b)) depicted in Figure 2: the rows 0 to 5 contain initial matchings of start node 0 from T. Row 6 contains matchings of all neighbors of node 0 from T with S. Node 0 has five neighbors (cf. Section 4.2) and instead of copying row 6 five times the matchings of neighbored nodes of node 0 have been appended to row 6.

4 Applications and Evaluation

In this section we will perform and evaluate two different tasks, the computation of maximal 3D building structures (Section 4.1) and the extraction of 3D building corners (Section 4.2). Both tasks are reasoned by the intensive use of building model knowledge during automatic extraction of man-made objects from images (Grün et al. 1995).

The applications are based on 3D building descriptions of the area Oedekoven close to Bonn consisting of 1846 building cluster (short: buildings). The buildings have been modeled three-dimensionally from aerial images with a high degree of detail (Englert and Gülch 1996) which includes e.g. eaves, detailed roof structures (cf. Figure 3). 745 buildings have a complex roof structure (neither flat- nor lean-to roof) and hence they are of major interest for automatic extraction. The buildings can be grouped into three classes: (a) detached houses (489), (b) any combination of two basic building types (162); cf. Figure 4, e.g. a L-shaped hip- and saddleback roof building, and (c) buildings which consist of more than two basic building types (94). Note, in most cases a combination of basic building types mean that some of the buildings are nested or closely interlocked resulting in a complex roof structure and in an irregular ground plane. Hence, without background knowledge about roof details the automatic extraction of roofs will be incomplete due to the complexity of the data set (Bignone et al. 1996).

From all buildings the surfaces have been computed (Englert 1997), and have been converted subsequently to the following p-graphs: parameter vectors of the nodes V are $\zeta(n \in V) = (D, G, CZD)$ with D is the degree of node n, G the gutter height of the corresponding building corner, and CZD is the value of the z-dimension of

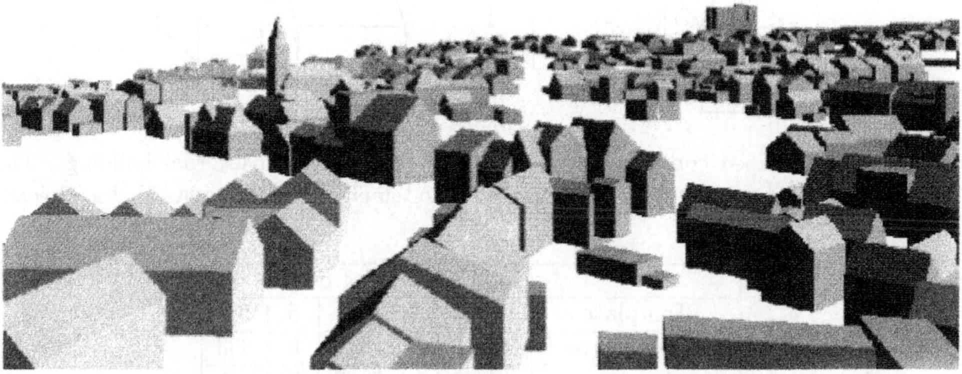

Fig. 3: A visualized part of the sample test field Oedekoven.

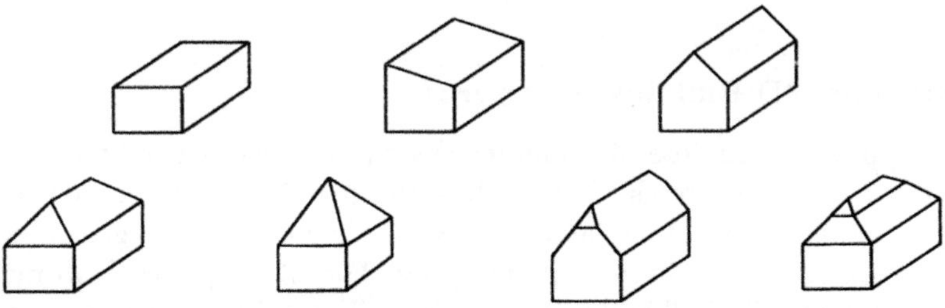

Fig. 4: Some basic building types with different roof shapes: flat roof, pent roof (lean-to roof), saddleback roof (gable roof), hip roof, broach roof (pavilion roof), hipped-gable roof, mansard roof (gambrel roof).

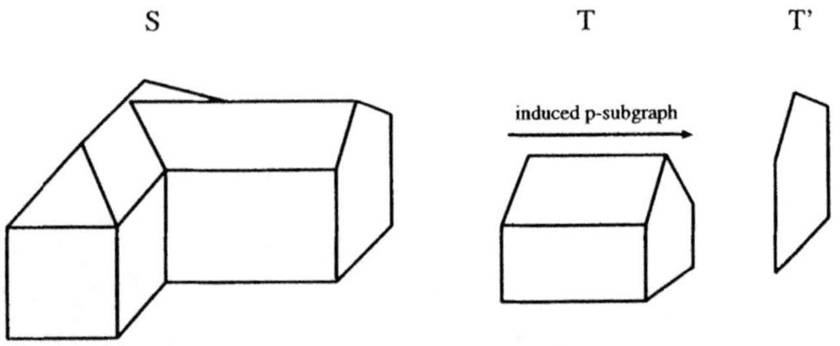

Fig. 5: A L-shaped combination of a hip-roof- and a saddleback roof building. The maximal induced p-subgraph T' of S according to template T is the right-sided gable-end.

	Number of nodes					
Template	1	2	3	4	5	6
Saddelback roof	0	24	0	1044	0	893
Hip-roof	0	60	0	7322	28	56

Tab. 1: Results of maximal induced p-subgraph computation using the templates saddleback roof and hip roof.

its object coordinate space, and parameter vectors of the edges E are $\xi(e \in E) = (\triangle, L, AHE)$, where $\triangle$ is the gradient of e, L the length of e, and AHE denotes if an adjacent orthogonal edge e' of e exists which is horizontal. The latter geometrical information is required in order to distinguish a hip-roof and a saddleback roof (cf. Figure 4).

4.1 Maximal 3D Building Structures

The first application explores 3D building descriptions in order to find most frequent occuring parts of given roofs. For guiding the search of aforementioned so-called *template p-graphs* (short: templates) are used. They denote background knowledge which is given as roofs of basic building types. For all templates T all maximal induced p-subgraphs T' of all building p-graphs S (in total 1846) are computed, where 'maximal' refers to the number of nodes of T, and grouped according to their size. Let us highlit an example: Figure 5 shows on the left side a L-shaped combination of a hip-roof- and a saddleback roof building of which p-graph S represents the search space. Using as template T a saddleback roof, the maximal induced p-subgraph T' of S according to T is the right-sided gable-end. The parameter vectors of the templates used here are depicted in figure 6, where the symbols X_i and Y_j denote variables, and for the variable Y_j must hold $]0.0, \ldots, 90.0[$.[4]
The frequency of occurrence of p_n-connected subgraphs with n nodes of the used templates are depicted in Table 1. The results conclude that the p-subgraphs of the

[4] $] \ldots [$ denotes an interval without its delimiting points.

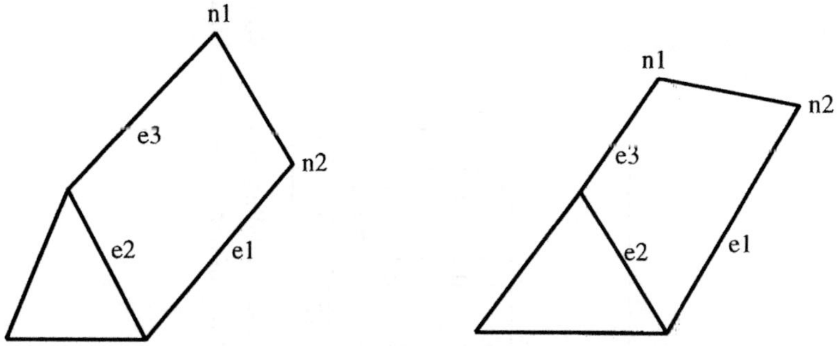

n1 = (3, X1, X2)
n2 = (3, X3, X4)

e1 = (0, X5, 0)
e2 = (Y1, X6, 1)
e3 = (0, X7, 0)

n1 = (3, X2, X9)
n2 = (3, X10, X11)

e1 = (0, X12, 0)
e2 = (Y2, X13, 0)
e3 = (0, X14, 0)

Fig. 6: The templates saddleback roof (left side) and hip-roof (right side) which serve as background knowledge in order to guide the search during maximal induced p-subgraph isomorphism computation.

appropriate templates with four nodes occur most frequent. Most extraordinary is the frequency of the recurring results in total 893 times with regard to the computed p-isomorphic graphs of the saddleback roof template. In conclusion for automatic building extraction from images these are the most important structures serving as building models.

4.2 3D Building Corners

The second application is to compute statistical information about various kinds of 3D building corners containing exactly three line segments. These structures are very important in automatic extraction of buildings from images, since they are the most frequent observable substructures of buildings (Gülch 1992). As p-graphs they are represented using four nodes (cf. Figure 7): one node, the so-called *corner-node* has three adjacent nodes and three edge labels out of horizontal (h), vertical$^+$ (v^+), vertical$^-$ (v^-), oblique$^+$ (o^+), or oblique$^-$ (o^-). The signs '+' and '−' denote the gradient of the line segment considering the corner-node. Figure 7 shows as example the corner nodes of a saddleback- and a hip-roof building. The task is twofold, (a) to extract all possible corner nodes of buildings which are reflected by Table 2 using the various types of corner nodes as templates (21 in total), and (b) to determine their frequency of occurrence. Note, the classes of corner nodes depicted in Table 2 are based on a classification according to a given error model (Englert and Cremers 1997).

Edge label	Frequency of edge labels																				
v^+	1	0	1	0	1	0	0	1	0	0	0	1	0	0	0	1	0	0	0	0	0
v^-	0	1	0	1	0	1	0	0	1	0	0	0	1	0	0	0	1	0	0	0	0
h	1	1	1	1	0	0	1	2	2	2	2	0	0	1	0	0	0	1	0	0	0
o^+	1	1	0	0	1	1	1	0	0	1	0	2	2	2	2	0	0	0	1	3	0
o^-	0	0	1	1	1	1	1	0	0	0	1	0	0	0	1	2	2	2	2	0	3
f_O	4	10069	2	25	0	2	5	21672	12710	807	55	0	14	63	3	0	0	4775	4	0	13

Tab. 2: The various types of corner nodes (21 in total) and their frequency of occurrence f_O (out of 1846 building cluster). Note, the total value of each column must sum up to 3. Those combinations of edge labels which induce structures of non three-dimensional nature must be ignored.

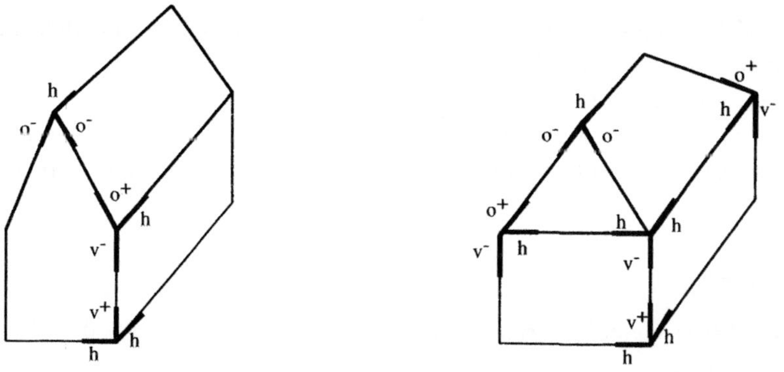

Fig. 7: Examples of corner nodes of a saddleback- and a hip-roof building with edge labels.

To resume Table 2 makes evident that there are certain categories: first, the corner nodes, e.g. (v^+, h, h), (v^-, h, h), and (v^-, h, o^+) occur most frequent. Second, the corner nodes, e.g. (h, h, o^-) or (h, o^+, o^+) occur seldom. Finally, some corner nodes, e.g. (v^+, o^-, o^-) do not occur at all. This histogram of occurrence of corner nodes gives numerical evidence for each corner-node and thus reasons their occurrence during automatic extraction of buildings from images.

5 Conclusion and Future Work

Parameterized graphs and induced p-subgraphs have been defined and an approach for subgraph isomorphism computation has been described. Background knowledge guides the systematic search of induced p-subgraphs. Two pruning mechanisms are used in order to reduce the search space. The approach has been applied to maximal p-subgraph isomorphism computation of search graphs by using template graphs resulting in significant parts of buildings, and 3D building corner computation and grouping according to their geometrical characteristics.

A challenging task is the following: often only parts of 3D building corners can be observed. Table 2 provides statistical knowledge for estimating missing parts of 3D building corners. Furthermore, the information gain given by the frequency of occurrence of 3D building corners can be estimated, too. These statistical analyses are depicted in (Englert and Cremers 1997). In addition, a grouping of computed maximal parts of roofs by using enhanced criteria will be instrumental to receive further insights of their characteristics and thus enable one to analyze the structure of buildings (Englert 1995).

Acknowledgments

The inspirations from Prof. Dr. Armin B. Cremers and Prof. Dr.-Ing. Wolfgang Förstner are gratefully acknowledged. The author likes also to thank E. Gülch

(Institute of Photogrammetry, Bonn), R. Fischer, and G. Gröger for their invaluable discussions.
This research is supported in part by BMBF/ DLR under Grant 01 M 3018 F/6. The views and conclusions in this document are those of the author and should not be interpreted as representing the official policies, expressed or implied of those agencies.

References

Aigner M. (1988) *Combinatorial Search*, John Wiley & Sons Ltd and B. G. Teubner, Stuttgart, Germany.

Besl P. J., R. C. Jain (1985) *Three-Dimensional Object Recognition*, Computing Surveys, Vol. 17, No. 1, pp. 75 – 145.

Bignone F., O. Henricsson, P. Fua, M. Stricker (1996) *Automatic Extraction of Generic House Roofs from High Resolution Aerial Imagery*, in B. Buxton, R. Cipolla (eds), Computer Vision - ECCV'96, Vol. 1 of Lecture Notes in Computer Science, Springer Verlag, pp. 85 – 96.

Braun C., T. H. Kolbe, F. Lang, W. Schickler, V. Steinhage, A. B. Cremers, W. Förstner, L. Plümer (1995) *Models for Photogrammetric Building Reconstruction*, Computer & Graphics, Vol. 19, No. 1, pp. 109 – 118.

Brunn A., E. Gülch, F. Lang, W. Förstner (1997) *A Multi-Layer Strategy for 3D Building Acquisition*, in F. Leberl (ed.), Proceedings of the IAPR-TC7 Workshop "Remote Sensing and Mapping", R. Oldenbourg, Wien, Austria, pp. 11 – 37.

Englert R. (1995) *Inducing Integrity Constraints from Knowledge Bases*, in I. Wachsmuth, C.-R. Rollinger, W. Brauer (eds), Proc. 19th Annual German Conference on Artificial Intelligence, Springer Verlag, Berlin, Germany, pp. 77 – 88.

Englert R. (1997) *Handling Irregularities of 3D Building Data During Surface Computation*, in N. Thalmann, V. Skála (eds), Proceedings of the Fifth International Conference in Central Europe on Computer Graphics and Visualization, Vol. I, University of West Bohemia Press, Plzen, CR, pp. 104 – 113.

Englert R., A. B. Cremers (1997) *Improving Reconstruction of Man-Made Objects from Sensor Images by Machine Learning*, in D. M. McKeown, C. McGlone, O. Jamet (eds), Proceedings of the 11th SPIE's International Symposium on AeroSense: Integrating Photogrammetric Techniques with Scene Analysis and Machine Vision III.

Englert R., E. Gülch (1996) *A One-Eye Stereo System for the Acquisition of Complex 3D-Building Structures*, GEO-INFORMATION-SYSTEMS, Journal for Spatial Information and Decision Making, Vol. 9, No. 4, pp. 16 – 21.

Faugeras O. (1993) *Three-Dimensional Computer Vision: A Geometric Viewpoint*, MIT Press, London, U.K.

Fontet M. (1976) *Linear Algorithms for Testing Isomorphisms of Planar Graphs*, in S. Michaelson, R. Milner (eds), Proc. 3rd Coll. Automata, Languages and Programming, Edinburgh University Press, Edinburgh, Scotland, pp. 411 – 423.

Fuchs C. (1996) *OEEPE-Study on 3D-City Models*, Workshop on 3D-City Models, Universität Bonn, Bonn, Germany, Institut für Photogrammetrie.

Grün A., O. Kübler, P. Agouris (eds) (1995) *Automatic Extraction of Man-Made Objects from Aerial and Space Images*, Birkhäuser Verlag, Basel, Switzerland.

Gülch E. (1992) *A Knowledge-based Approach to Reconstruct Buildings in Digital Aerial Imagery*, Proc. of 17th ISPRS Congress, Washington D. C., U.S.A., pp. 410 – 417.

Harary F. (1972) *Graph Theory*, Addison-Wesley, Massachusetts, U.S.A.

Hopcroft J., C. Wong (1974) *Linear Time Algorithms for Isomorphisms of Planar Graphs*, 6th Ann. ACM Symp. Theory of Computing, Seattle, U.S.A., pp. 172 – 184.

Mehlhorn K., S. Näher (1995) *LEDA, a Platform for Combinatorial and Geometric Computing*, Communications of the ACM, Vol. 38, No. 1, pp. 96 – 102.

Ullmann J. (1976) *An Algorithm for Subgraph Isomorphism Testing*, Journal of ACM, Vol. 23, pp. 31 – 42.

Semanitic Modeling for the Acquisition of Topografic
Information from Images and Maps
ed. by W. Förstner and L. Plümer

Building Recognition with Bayesian Networks

K. Kulschewski

Institut für Theoretische Geodäsie
Rheinische Friedrich–Wilhelms–Universität Bonn
Nußallee 17, 53115 Bonn, Germany
Ph.: +49-228-73 35 77, Fax: +49-228-73 30 29
e-mail: kai@theor.geod.uni-bonn.de

Abstract

This paper is concerned with a new approach to the recognition of types of buildings from digital aerial images. For this purpose, the technique of knowledge based image interpretation is used. The knowledge base is introduced via a geometrical and a semantical model. The interpretation is accomplished by a Bayesian network which allows to make decisions in systems with uncertainties. First, the theory of Bayesian networks is explained. For the recognition of buildings a Bayes net is developed in which the observed image features, the types of buildings and prior information are represented. Emphasis is laid on the description of the components of the network and their dependencies. Observable features are introduced into the net and by the propagation of probabilities the types of buildings are recognized. The observations are introduced sequentially but not in arbitrary order. Thus, criterions are derived which control the calculations. The performance of the Bayesian network is demonstrated by some examples.

1 Introduction

In the recent years an increasing demand exists for 3-D digital city models. Especially the users of geoinformation systems, for example the telecommunication industry, is interested in these data. For the planning of radio networks in inner city areas knowledge of the roof shapes is needed (Siebe and Büning 1996). The paper presented here deals with the localization of buildings in a digital image and especially with the recognition of the types buildings. Severals researchers have carried out approaches for an automatic extraction of buildings (Bignone *et al.* 1996; Lang and Förstner 1994).
The determination of the types of buildings is part of an automation process of the image interpretation. In (Braun *et al.* 1995) a model for the image interpretation

is presented. Starting with observed pixels, features are derived which characterize the objects existing in an image. From these features aspects, which are 2-D views of 3-D objects, are infered. Given the aspects it is possible to derive the label of the 3-D objects.

Especially the aspects will have an important status in this paper. Fischer and Steinhage (1996) deals with the modeling of the buildings' aspects and the determination of the aspects' frequencies. A reduction of the number of aspects by unification of identical aspects can be found in (Laurentini 1995). As one of the first Dickenson et al. (1992) used aspects for the transition of observable 2-D data to the third dimension. They started with the observations and generated hypotheses for the appearence of face primitives. By taking conditional probabilities into account, they where able to infer the aspects and thereby the 3-D primitives.

Results of low level digital image processes, i.e. segmentation and feature extraction, are prone to errors. The use of Bayesian networks allows decisions in a system with uncertainties. Here, knowledge is represented as random variables and conditional probabilities express the dependencies of the random variables. Due to this representation the knowledge is structured and the results of the modification of knowledge can be calculated. Essential contributions to the theory of Bayesian networks is given in (Pearl 1988; Neapolitan 1990).

The basic idea of the approach presented here is to use the Bayesian networks to formulate a model in which buildings can be recognized by calculation of probabilites. First of all the network has to be composed. Possibilities for an automatic design of a Bayesian network can be found in (Sarkar and Boyer 1994). If there are many random variables this can result in problems of the numerical handeling. Agosta (1990) gives proposals for the structure of the networks to reduce this problem.

In contrast to the usual forms of knowledge representation, i.e. semantic networks, conditional probabilities are needed in the Bayesian networks. Their acquisition is mostly difficult. Dickenson et al. (1992) and Mann and Binford (1994) present solutions of this problem.

If the network structure is modeled and the conditional probabilities are generated, the Bayes net can be used for the recognition of the objects. An application of Bayesian networks in the digital image processing can be found in (Rimey and Brown 1994). Moreover, Jian-Ming et al. (1994) applied the networks for the recognition of 3-D primitives.

The knowledge obtained from the digital image is not introduced into the net in arbitrary order. Thus, a control algorithm is needed for introducing the knowledge. In chapter 2 the theory of the Bayesian networks is explained. The third and the forth section deal with modeling of the network and the control of the recognition process. In last chapter the performance of the Bayesian networks is demonstrated by some examples.

2 Theory of Bayesian Networks

Bayesian networks belong to the class of expert systems, in which knowledge and their dependencies can be represented and processed. Knowledge is described in Bayesian networks as discrete random variables and their dependencies which exist among them. If there is a dependency between two random variables A and B and B depends on A, it can be expressed by the conditional probability $P(B|A)$. It is the probability that the existing knowledge A is relevant for the further information B. Introducing knowledge in a random variable A in form of the values of A, the probabilities of the values of B can be calculated by using the conditional probability $P(B|A)$. If further random variables depend on B their probabilities can be determined in the same way.

A Bayesian network is a directed acyclic graph, where the nodes are random variables. The arcs indicate the existence of direct causal influences between the linked variables, and the strenght of these influences are expressed by conditional probabilities (Pearl 1988, p.117). The equations for the calculation and the propagation of the probabilities in a Bayesian network are given by (Pearl 1988, p.183)

$$BEL(x_j) = \alpha \ \lambda(x_j) \ \pi(x_j) \tag{2.1}$$

where the vector BEL(x_j) denotes the probability of the random variable x_j under the condition that some evidence is given. The vector $\lambda(x_j)$ denotes the diagnostic support, the vector $\pi(x_j)$ the causal support and α the normalization constant. Furthermore

$$\pi(x_j) = \sum_{u_1,..,u_n} p(x_j|u_1,..,u_n) \prod_i \pi_{x_j}(u_i), \tag{2.1a}$$

$$\lambda(u_i) = \prod_j \lambda_{x_j}(u_i), \tag{2.1b}$$

$$\pi_{x_j}(u_i) = \alpha \frac{BEL(u_i)}{\lambda_{x_j}(u_i)}, \tag{2.1c}$$

$$\lambda_{x_j}(u_i) = \beta \sum_{x_j} \lambda(x_j) \sum_{u_k:k\neq i} p(x_j|u_1,...,u_n) \prod \pi_{x_j}(u_k), \tag{2.1d}$$

where x_j denotes the variables depending on u_j.

As an example let the Bayesian network shown in Figure 1 be given. The nodes represent the random variables whose probabilities can be derived and the conditional probabilities are pictured as arrows between the nodes. From (2.1) we see, that the probability of random variable x_4 is composed of two parts. The first one stems from the knowledge propagated in direction of the arrows (from the parents to the childrean) to x_4 which is called $\pi(x_4)$. The second one is $\lambda(x_4)$ that propagates

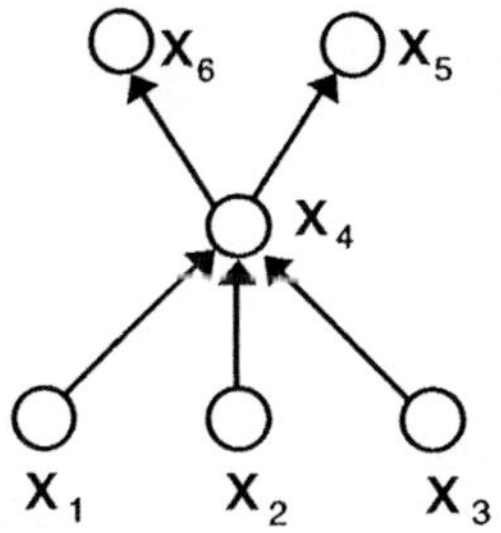

Fig. 1: Bayes net

information from all nodes that depend on x_4 against the arrows (from the children to the parent). The belief is obtained from (2.1) as

$$BEL(x_4) = \alpha \; \lambda(x_4) \; \pi(x_4). \tag{2.2}$$

As mentioned the constant α normalizes the equation. Thus, the second axiom of probability is fulfilled

$$\sum_{x_4} BEL(x_4) = 1.$$

At the beginning of the calculations there is no knowledge existing. Hence every $\lambda(x_i)$ and $\lambda_{x_j}(x_i)$ are postulated as equally distributed. It means that we have to set $\lambda(x_i) = \lambda_{x_j}(x_i) = \alpha(1, 1, ..., 1)$. If knowledge is introduced into the net we can distinguish two different types of knowledge. On the one hand there exists prior knowledge. This is introduced by known distributions for some random variables. On the other hand we have observations of different random variables in form of their density functions. By introducing these evidences the resulting probabilities can be propagated through the Bayes net. This gives changes of the density functions of the random variables and thereby modified probabilities.

In the Bayes net pictured in Figure 1 the conditional probabilities are assumed to be given and the initialization has been completed. From prior information the distributions $\lambda(x_5)$ and $\lambda(x_6)$ of the random variables x_5 and x_6 are known. This prior knowledge is propagated using (2.1d), (2.1b) and (2.1)

$$\begin{aligned} \lambda_{x_i}(x_4) &= \beta \sum_{x_i} \lambda(x_i) p(x_i|x_4) \pi_{x_i}(x_4) \quad , i \in \{5, 6\}, \\ \lambda(x_4) &= \lambda_{x_5}(x_4) \lambda_{x_6}(x_4), \\ BEL(x_4) &= \alpha \lambda(x_4) \pi(x_4), \end{aligned}$$

into the node x_4, where the modified knowledge BEL(x_4) and $\lambda(x_4)$ are obtained. By additional application of (2.1d), (2.1b) and (2.1) the belief vectors of the random variables x_1, x_2 and x_3 can be derived. The way of propagation is shown in Figure 2. If an observation in form of a density function is executed, for example $\pi_{x_4}(x_1)$,

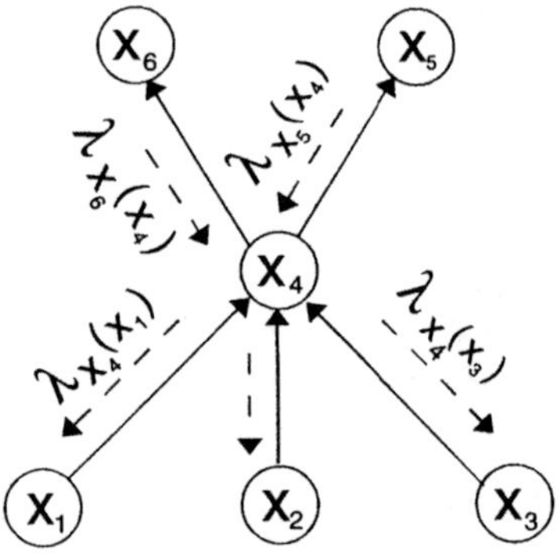

Fig. 2: Propagation of prior knowledge

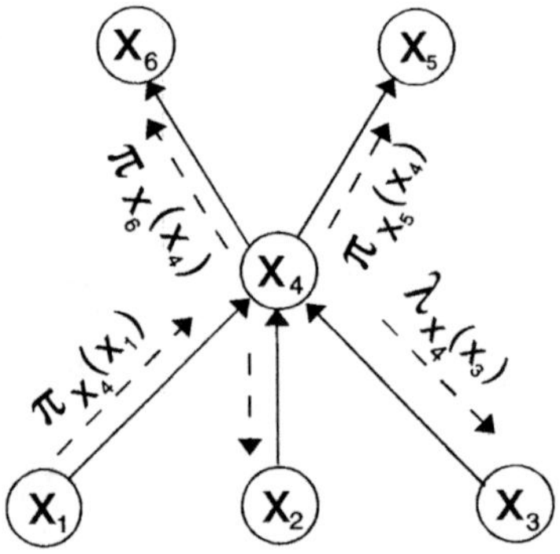

Fig. 3: Propagation of observed information

it can be implemented into the net. Using (2.1c), (2.1a) and (2.1) we obtain

$$\begin{aligned}
\pi_{x_4}(x_i) &= \alpha \frac{BEL(x_i)}{\lambda_{x_4}(x_i)} \quad ,i \in \{1,2,3\}, \\
\pi(x_4) &= \sum_{x_1}\sum_{x_2}\sum_{x_3} p(x_4|x_1,x_2,x_3)\pi_{x_4}(x_1)\pi_{x_4}(x_2)\pi_{x_4}(x_3) \\
BEL(x_4) &= \alpha\lambda(x_4)\pi(x_4),
\end{aligned}$$

BEL(x_4), the modified knowledge in x_4, is calculated and by further propagations (see Figure 3) with the same equations we receive BEL(x_5) and BEL(x_6). On the other hand the observations introduced yield the changes of the knowledge in x_2 and x_3. This can be calculated with (2.1d), (2.1b) and (2.1). A different method of calculating the probabilities in a Bayesian network is given by (Koch 1995). It has been used to check the formulars which have been derived.
In the next chapter a Bayes net for the recognition of buildings is deviced. In chapter 4 we make use of the equations (2.1) and (2.1a-d) in the Bayes net derived.

3 A Bayesian Network for Recognition of Buildings

The digital image processing generally consists of two levels. At the lower level a segmentation of an image in faces, edges and points is performed. In an additional step geometrical, topological and semantical features are determined for those objects which have to be identified at an upper level of the image processing. Both the segmentation and the feature determination are prone to errors. In order to handle these uncertainties, a Bayesian network is used for the recognition. In the following a model is developed which recognizes different types of buildings in a scene based on information derived from a digital image. The components of the net, the dependencies of the nodes, the prior knowledge and the generation of the conditional probabilities are explained. The Bayes net which has been developed is shown in

Figure 8 at the end of this chapter.
In general, buildings are very complex objects. Due to the architects' endeavour to create unique houses it is nearly impossible to recognize every type of building if the process of recognition is based on geometrical and relational object features. But the major part of existing buildings can be reduced to the types of buildings shown in Figure 4.

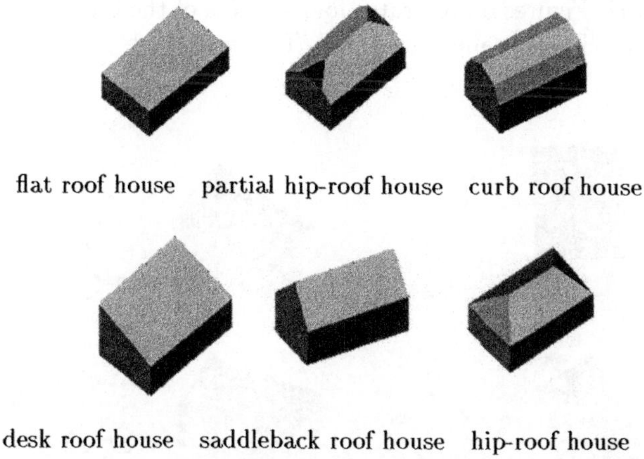

Fig. 4: Types of buildings

We will assume that either these types occur or a fusion of these types form a building. This yields the first node of the Bayes net, which is called *building (g)*, containing the six types of buildings as values of this random variable. In the next step projections of the 3-D objects into 2-D are introduced. An isomorphical view of a 3-D Object in 2-D is called an aspect (Steinhage 1995). In an isomorphical view the same faces, edges and corners are visible in an identical neighborhood structure. To reduce the complexity of calculations equal aspects have to be united, leading to the question of how equal aspects can be defined. In the following it is assumed that two or more aspects are equal if they have the same number of faces, equal face adjacency graphs and corresponding characterization of the surfaces. For example

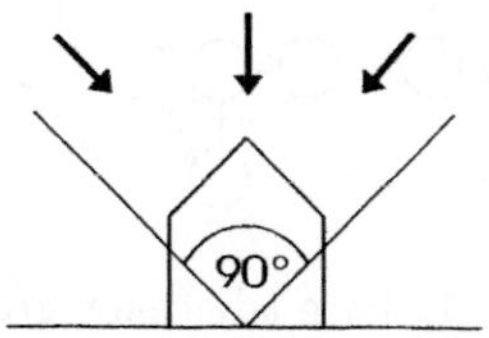

Fig. 5: Cone of view

the view of a flat roof house such that only the roof face and one rectangular wall face are visible can be taken from four different points of view. If the fact, that in each aspect different wall faces are visible, is neglected, the four aspects are postulated to be equal which results in an unification.
A further reduction of the number of aspects occurs by taking only those views into account which form a cone with an aperture angle of 90 degrees (see Figure 5). This restriction has to be introduced, when analyzing aerial photographs. It leads to 25 different aspects of the six types of buildings, which label the values of the random variable *aspect (a)*. Figures 6 illustrates four aspects of the hip roof house where '3' denotes a triangle, 'P' a parallelogram and 'T' a trapezoid.

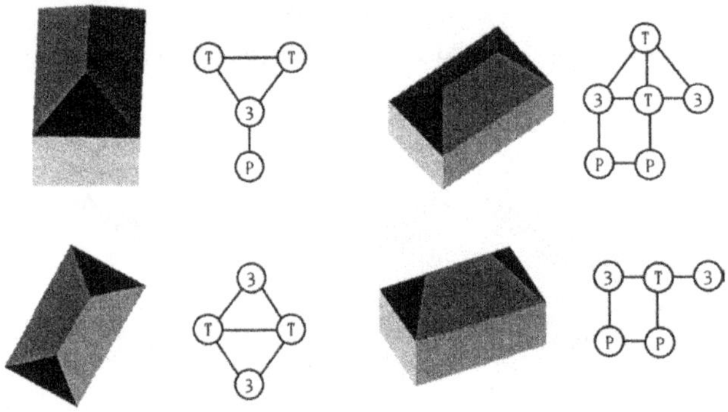

Fig. 6: Aspects of a hip roof house

In the next lower level of the Bayes net features are introduced which characterize the aspects. Therefore the three criterions mentioned above for distinguishing between the different aspects are specified. By introducing a random variable *neighborhood (N)* it is possible to represent the number of faces and their face adjacency graphs. The 23 values of this random variable are the numbers of the face adjacency graphs, which appear in the given aspects. The Figure below shows ten of them.

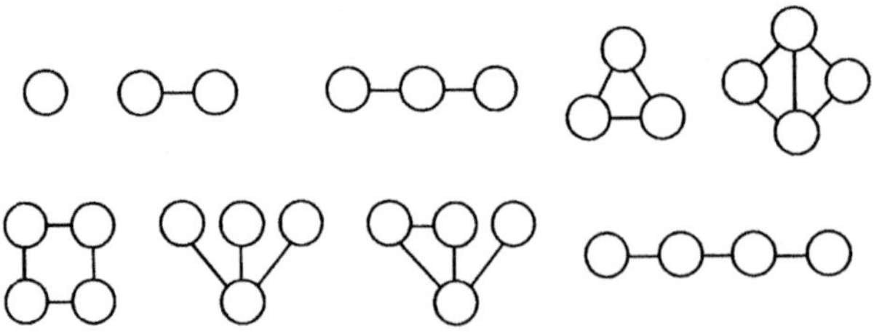

Fig. 7: Face adjacency graphs

The node *aspect (a)* depends on the random variable *neighborhood (N)*. Observations

and the resulting instantiations of the face adjacency graph can be used to infer the presented aspect by taking the conditional probabilities $P(a|N)$ into account.
Some basic neighborhood structures appear in a number of aspects. To capture the resulting ambiguities, the aspects have to be additionally characterized. The shapes of the faces are introduced as attributes of the surfaces. The given aspects contain six different shapes of faces: triangle, parallelogram, trapezoid, pentagon, hexagon and heptagon. Thus, the random variable *face* (f_i) gets six values.
So far each node was implemented only once in the Bayes net. However, the number of the random variables *face* (f_i) depends on the number of faces of the particular aspects. To give an illustration, there are four nodes f_1, f_2, f_3 and f_4 needed for the recognition of upper left aspect in Figure 6 whereas six random variables are necessary for the upper right one. Thus, the Bayes net is constructed dynamically and changes its structure with the number of faces introduced as observations.
The shapes of the faces are described by geometrical features like symmetrical axes and parallel edges. Futhermore, semantical descriptions like 'Is a face part of a roof ?' are used. For each feature a random variable *feature* (l_i) is introduced.
To complete the Bayes net, additional prior information is introduced and the dependencies of the nodes are explained in the following. It is obvious that the types of buildings do not appear equally likely. Hence, a node called *frequencies of buildings* (h_g) is introduced to represent the probabilities of the appearance of each building. The frequencies depend on the local area and the year of construction.
In the same way the aspects are not equally distributed. But the probabilities of appearance can be derived analytically (Fischer and Steinhage 1996) and implemented for the random variable *frequencies of aspects* (h_a). Furthermore, the probabilities of covered as well as non available faces are determined. This leads to the node *covering* (V). Details will be discussed in Chapter 4.

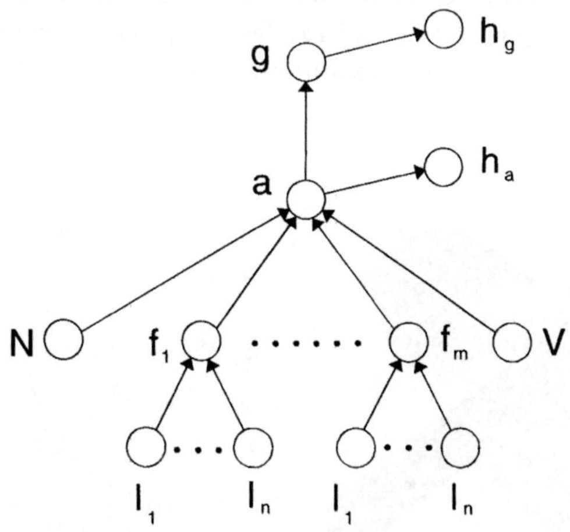

Fig. 8: Bayes net for building recognition

After all nodes of the Bayesian network are explained their dependencies have to be

specified. It is assumed that the conditional dependencies are known. Starting with observed features it is possible to identify the faces and by taking the neighborhood structure additionally into account, the aspects and finally the types of buildings are infered. Generally, the probabilities of the values of the random variables in an upper level are determined, given the knowledge of the lower level. This procedure yields to the dependencies of the nodes automatically. The three prior nodes are introduced as dummy nodes (Pearl 1988), which results in reverse dependencies.
For the calculations within the Bayes net the conditional probabilities have to generated. Given the aspects, the appearance of the types of buildings can be denoted and weighted with the probabilities of the aspects' appearance. The normalization of these relative frequencies yields the conditional probabilities $P(g|a)$. For each aspect the occurrence of the types of faces and their neighborhood are determined whereby the conditional probabilities $P(a|f_1, ..., f_i, N)$ are resulting. The other conditional probabilities can be generated in the same manner.
Thus, the complete data except the observations N, l_i is given for the recognition of buildings with the Bayesian network shown in Figure 8. The two subsequent chapters deal with the recognition algorithm and the practical usage of the developed Bayes net, respectively.

4 Control Structure for Recognition

It will be assumed in the following that the lower level of the digital image processing supplies regions and their neighborhood. A face adjacency graph is constructed which represents the topology of the 2-D image components. In addition, edges, points and their relation to the regions are known from the segmentation. By the assignment of edges and regions and with further information, for instance a digital elevation model, the features of the faces are derived. As an example Figure 9 shows a segmentated object scene and his face adjacency graph, while Figure 10 only presents the graph and the corresponding labels of the nodes. The task of the

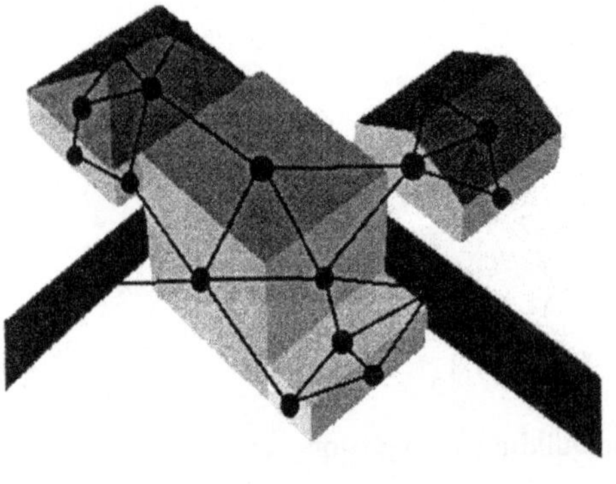

Fig. 9: Object scene

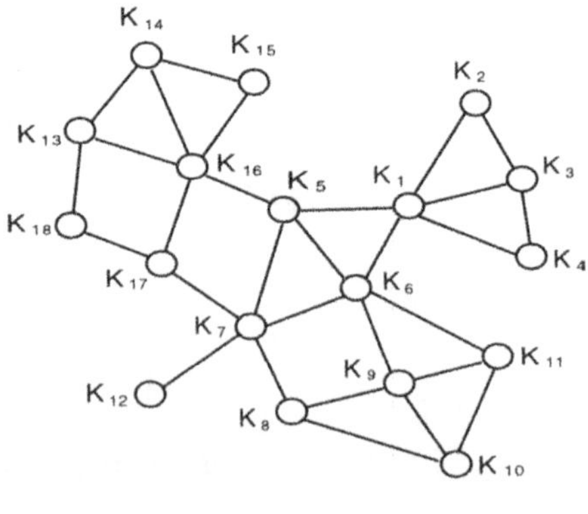

Fig. 10: Face adjacency graph

Bayesian network is to combine the faces to an aspect from which a type of building follows. Generally, there exists more than one object in a scene yielding the problems that it is not known where buildings appear and which faces are part of them. Starting the calculations with one face, additional faces have to be introduced successively into the Bayesian network and the propagation of probabilities has to performed. This raises the following questions, which are answered in this section.

- Which face has to be introduced into the Bayes net first ?
- By which order are the additional surfaces added ?
- How to decide whether a face, which has been introduced, is part of an aspect ?
- When does the procedure end ?

There are three criterions that can help to find the starting face. First, a digital elevation model can give information about the position of roof faces. By using this information, the search area for the buildings can be reduced. Thus, a roof face has to be introduced into the Bayes net first. Second, significant surfaces, for instance node K_1 in Figure 10, could be the starting face for the recognition. It is also possible to start with the face with maximum belief. The probabilities of all values of the random variables *face* are calculated by the observed features using (2.1c), (2.1a) and (2.1) yielding $BEL(K_c)$, where c denotes the nodes. The face K_b with maximum belief is given by the MAP estimate $maxBEL(K_b)$.
Because the first mentioned criterion is the best one to start the process with, the others are only used if different roof faces can be chosen. After selecting the starting surface the information is entered into the net. Next the node *neighborhood* has to be instatiated. This yields a first propagation and calculation of the probabilities in the node *building*.
With the assumption that there are no coverings that split a building into two parts, other faces corresponding to the same aspect can be found in the neighborhood of the first face. Thus, from the set of the neighbors the second and all following nodes are chosen. Again, different criterions for selecting the next face have to be used. On the first hand the information of each face can be introduced into the Bayes net and the one which produces the highest increase of probability in the node *building* by propagation is taken to cluster it with the former nodes. An alternative to this expensive method is the prediction of the observation. Here the former observations are propagated into a fictitious random variable face f_f using (2.1) - (2.1d). The resulting $BEL(f_f)$ is interpretated as follows: If a face of the type with the highest probability in $BEL(f_f)$ is introduced into the net, it will yield the highest increase of probability in the node *aspect* or in the node *building*, respectively. Then, if $BEL(K_c)$ are the probabilities of the values of all neighborhood faces, $BEL(K_c)$-$BEL(f_f)$ has to be minimized. Because the difference does not consider a high $BEL(K_c)$, the formular has to be modified. The node K_k fulfilling

$$max \quad V(K_k) = \frac{BEL(K_c)}{exp(BEL(K_c) - BEL(f_f))} \tag{4.1}$$

is chosen to be implemented into the Bayes net. If it is known which 2-D adjacent faces are not neighbors in 3-D, the search area for the introduced face can be reduced. In 3-D surfaces are called neighbors if they have at least one common edge in a way that both ends of this edge in face K_1 correspond to both ends in face K_2. By using this condition in 2-D the result that two faces, for instance K_1 and K_5, belong to the same building can be prevented. But this is not a sufficient criterion, as can be seen by taking a look at the nodes K_1 and K_5. They are parts of the same building but do not satisfy the condition. The choice of the second and the following nodes take place by the use of both criterions.
The successive introduction of faces into the net is repeated until no face can be implemented which yields an increasing probability in the node *building*. The clustered faces are deleted from the face adjacency graph and the process of recognition is continued until all nodes are labeled with a type of building or declared as unknown.
In the following an exemplary interpretation of the calculations in the Bayes net is given. Hereto it is assumed that the nodes K_1 and K_2 are recognized as rectangle roof faces and have been introduced into the Bayes net. In the random variable *aspect* the probability of the appearance of the different aspects is calculated, given two neighbored rectangle roof faces. One can see that only the probabilities of aspects of the saddleback roof house and the probabilities of aspects of the curb roof house are unequal to zero. If the prior information of the probability of the aspects' appearance is not taken into account this leads to equal probabilities of these aspects. The fact that the aspects of the curb roof house are built of much more than two faces remains unconsidered. But if these certain aspects are present there have to be not only two rectangle roof faces but also at least faces which are missing. So the probability of that situation is lower than for simpler aspects. The reduction of the probability is realized by the introduction of node *covering (V)*.
Based on the derived Bayes net from chapter 3 and the control structure explained in this section some examples of the recognition of building are given in the following.

5 Results

As already mentioned, given the six types of buildings there exist 25 different aspects. In a first example one aspect after the other is used as a test object. In a second example the developed Bayes net is utilized in the object scene shown in Figure 9.

Example 1: Using all aspects as observations allows to verify the derived Bayes net because the net itself was built up on these very aspects. In the best case each observation will give the result that the corresponding building and aspect are recognized with a probability of 1. Each object scene contains only one aspect. Hence, no other objects and coverings are presented. The actual 25 calculations show that in 22 cases the buildings are recognized with the maximal probability. The reason is the missing of uncertainties and the uniqueness of relation of aspects and buildings. In the left three cases there exists more than one hypothesis for the building label so that only two buildings were recognized. If two rectangle faces are visible this aspect can either be a flat roof house, a desk roof house, a curb roof house or a saddle-back roof house.

We now focus our attention on the aspects' probabilities. If *aspect no. i* has to be recognized the probability of the i-th value of the random variable *aspect* should be the largest one. For each aspect the corresponding probability after introducing all aspects' faces is pictured in Figure 11. If there are no ambiguities the probabilities

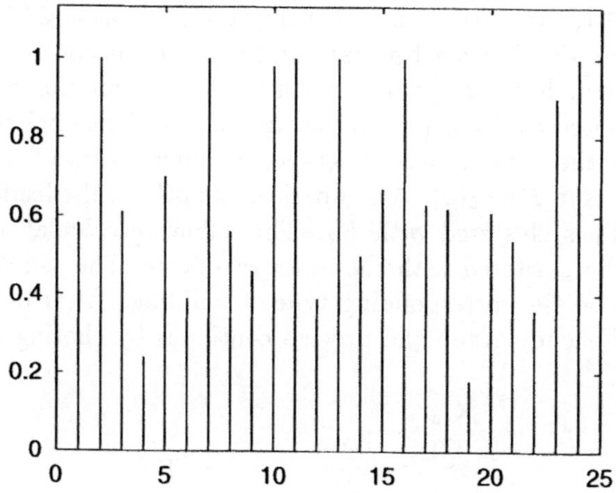

Fig. 11: Probabilites of each aspect

are supposed to be one. This is not always true because simpler aspects can be found within more complicate aspects. The aspect pictured in Figure 6, upper left case, corresponds to aspect number 18 in Figure 11, the upper right one to number 24, the lower left to 19 and the last one to 20.

If all observations of aspect no. 19 are introduced, this aspect is recognized with a probability of only 0.18. The reason is that its faces and the neighborhood structure appear in a lot of other aspects, too, so an unique classification is not possible. If the recognition process is done for aspect no. 20, the 20-th value of BEL(a) is 0.61 and the 24-th is 0.39. Hence a unique classification of the aspect fails, too, but the labeling with the type of building is correct, because both aspects correspond to the hip roof house. The most complicated aspect no. 24 is recognized with a probability of one. Considering these results it can be noticed that the major part of the aspects are recognized with probability higher than 0.5. The probabilities will increase if further features of the aspects are introduced into the Bayes net. Anyway, the labeling of the buildings is nearly unique in this example, leading to the result that the use of the developed Bayes net in the recognition process makes sense.

Example 2: Here, the object scene shown in Figure 9 is to be analysed. The presented four buildings have to be localized and recognized. The differences between this and the former example are:

- more than one building is presented,
- there are also non-building faces and
- buildings are partly covered.

Starting with the observations of the faces' features, the probabilities of the faces BEL(K_i) with i $\in$ 1,..,18 are determined using (2.1c), (2.1a) and (2.1). The roof face K_2 is chosen to be the first starting face and the information is propagated in the Bayes net into the node *building* which yields the first calculation of the probabilities in this node, denoted as $BEL_1(g)$. By using the criterions explained in chapter 4, K_3 is selected as the second face, which is implemented into the Bayes net. The propagation results in $BEL_2(g)$. A comparison of both results leads to an increasing probability, so K_3 is clustered with K_2. The following calculations show that K_1 and K_4 have to be clustered with the other two faces. The saddleback roof house is determined to be the corresponding type of building. After deleting these faces from the face adjacency graph the process continues by chosing the next starting

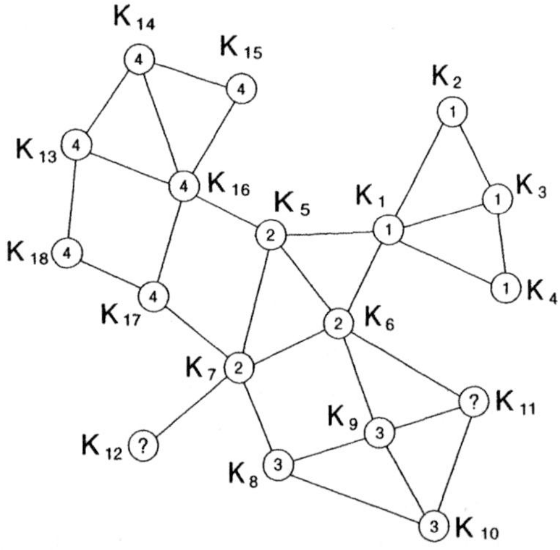

Fig. 12: Results of the second example

face. The result of the whole recognition process is shown in Figure 12. Numbers indicate the membership of the faces to an aspect. Aspect 1 is labeled as saddleback roof house, aspects 2 and 3 are labeled as flat roof houses and aspect 4 is labeled as hip roof house, while the nodes K_{11} and K_{12} are determined to be unknown.
Up to now the Bayes net was used with generated data. If we have to deal with real

data, the construction of the Bayes net has to be expanded to respresent the additional uncertainties of the observations. Furthermore, an introduction of semantical information, e.g. the appearence of streets, make it easier to localize the buildings. Especially the neighborhood of streets is a strong criterion for appearance of buildings.
The developed method shows, that the use of Bayesian networks in the process of the buildings' recognition is promising. This can be explained by the great flexibility of the net structure.

6 Acknowledgements

The support of the Deutsche Forschungsgemeinschaft (DFG) is gratefully mentioned.

References

Agosta, J.M. (1990) The structure of Bayes networks for visual recognition. In: Shachter, R.D., T.S. Levitt, L.N. Kanal and J.F. Lemmer (Eds.), *Uncertainty in Artificial Intelligence, Vol. 4*. North Holland, Amsterdam, 397–405.

Bignone, F., O. Henricsson, P. Fua and M. Stricker (1996) Automatic extraction of generic house roofs from high resolution aerial imagery. In: Buxton, B. and R. Cipolla (Eds.), *Computer Vision - ECCV'96, Vol.I.* Springer, Berlin, 85–96.

Braun, C., T. Kolbe, F. Lang, W. Schickler, V. Steinhage, A.B. Cremers, W. Förstner and L. Plümer (1995) Models for photogrammetric building reconstruction. *Computer and Graphics*, 19:109–118.

Dickinson, S.J., A.P. Pentland and A. Rosenfeld (1992) 3-D shape recovery using distributed aspect matching. *IEEE Trans Pattern Anal Machine Intell*, 14:174–198.

Fischer, A. and V. Steinhage (1996) *On the Computation of Visual Events in Aspect Graph Generation.* submitted paper.

Jian-Ming, L., H.I. Christensen and F.V. Jensen (1994) Qualitive recognition using Bayesian reasoning. *Proc of Pattern Recognition in Practice IV.* URL ftp://ftp.iesd.auc.dk/pub/reports/liang-christensen:jensen:94.ps.gz.

Koch, K.R. (1995) Bildinterpretation mit Hilfe eines Bayes–Netzes. *Z Vermessungswesen*, 120:277–285.

Lang, F. and W. Förstner (1994) Surface reconstruction of man–made objects using polymorphic mid–level features and generic scene knowledge. *Z Photogrammetrie und Fernerkundung*, 6:193–202.

Laurentini, A. (1995) Introducing the reduced aspect graph. *Pattern Recognition Letters*, 16:43–48.

Mann, W.B. and T.O. Binford (1994) Probabilities for Bayesian Networks. *Proceedings Image Understanding Workshop 1994*, Morgan Kaufmann, San Francisco.

Neapolitan, R.E. (1990) *Probabilistic Reasoning in Expert Systems.* Wiley, New York.

Pearl, J. (1988) *Probabilistic Reasoning in Intelligent Systems.* Morgan Kaufmann, San Mateo.

Rimey, R.D. and C.M. Brown (1994) Control of selective perception using Bayes nets and decision theory. *Int J Computer Vision*, 12:173–207.

Sarkar, S. and K.L. Boyer (1994) Automated design of Bayesian Perceptual inference networks. *Proc 1994 IEEE Comp Soc Conf Comp Vision Pattern Recognition*, IEEE Computer Society Press, Los Alamitos, 98–100.

Siebe, E. and U. Büning (1996) 3D–Stadtmodelle für Funknetzplanungen im Mobilfunkbereich. *GEO–Informations–Systeme*, 4:21–30.

Steinhage, V. (1995) Generische CSG–Modellierung und Aspekthierarchien zur Gebäudeerkennung in digitalen Luftbildern. In: *15. Jahrestagung der Deutschen Gesellschaft für Photogrammetrie und Fernerkundung, Hannover, 4.-6. Okt. 1995, Band 4 der Publikationen der DGPF.*

Semanitic Modeling for the Acquisition of Topografic
Information from Images and Maps
ed. by W. Förstner and L. Plümer

Model-Based Surface Recovery of Buildings in Optical and Ranges Images

Christopher O. Jaynes, Allen Hanson, and Edward Riseman

Computer Vision Laboratory
University of Massachusetts
Dept. of Computer Science
Box 34610, University of Massachusetts
Amherst, MA. 01003
e-mail: jaynes@cs.umass.edu, url: http://vis-www.cs.umass.edu

Abstract

A technique is introduced for extracting and reconstructing a wide class of building surface types from a registered range image and optical image. An attentional focus stage, followed by indexing into a library of surface models based on the differential geometry of a region, allows top-down robust surface fittting to reconstruct the 3D nature of the buildings in the data. Because of the effectiveness of model selection, top-down processing of noisy range data still succeeds and the algorithm is capable of detecting and reconstructing several different building roof classes, including flat single level, flat multi-leveled, peaked, and curved rooftops.

The algorithm is applicable to range data that may have been collected from several different range sensor types. Our results underline the usefuless of range data when processed in the context of a focus-of-attention area derived from the monocular optical image. We demonstrate the completely automatic reconstruction of buildings in the presence of large amounts of noise using the techniques described.

1 Introduction

We are concerned with the extraction of building surfaces from both range and optical images and have developed techniques that are are robust with respect to noise in both the optical and range images and are applicable to a wide range of building classes. The reconstructed models may be used for urban planning, three dimensional visualization for simulated walk/fly throughs, and as a geometric representation capable of supporting additional image understanding tasks.

The detection and accurate reconstruction of three dimensional scene structures from aerial data proves to be difficult in both optical and range data. Within optical imagery, building boundaries are occluded by neighboring buildings, trees, shadows, and other features at the site. Rooftop regions contain varying surface structure such as air vents and may be cluttered with surface markings. Problems in range images are similarly complex. Depth discontinuities at the boundaries of buildings cause typical correlation-based stereo optical systems to degrade. Likewise, depth computed from radar may be noisy and depend on surface geometry and material properties of objects in the scene.

A successful automated reconstruction system will make use of both range and optical data in order to overcome the inherent problems in each. The system should be robust with respect to a wide vareity of range sensors including laser altimeters, Interferometric Synthetic Aperature Radar (IFSAR), and optical stereo. This paper introduces one such system that demonstrates how appropriate use of both the registered optical image and range data increases reconstruction accuracy, reliability and completeness, even when applied near the feasible limits of sensor resolution.

There has been a large amount of work in aerial image interpretation, and building reconstruction in particular, using both monocular and multiple image strategies. (Collins'96, Jaynes'94, Herman'94, Matsuyama'85) This paper presents significant progress through the fusion of elevation data and optical imagery, the use of available domain knowledge to constrain the reconstruction process, and the development of valuable context-sensitive techniques for the aerial image domain.

In earlier work, perceptual organization techniques provided the impetus for many building extraction systems. Heurtas and Nevatia (Huertas'80), organize lines and corners into possible rectangles and select the best possible set of groupings from the hypotheses. The ASCENDER system (Collins'95) hypothesized 2D building rooftops through a perceptual grouping scheme (Jaynes'94) and computed a height estimate for each roof through multi-image triangulation. The system assumed flat roofed buildings and simply extruded the polygon to the assumed known ground plane to arrive at a reconstruction.

More recently, the utility of range images for site reconstruction has been recognized. Kim and Muller (Kim'95) extract rooftop boundaries from optical data and use an elevation map to estimate rooftop height. Haala and Hahn (Haala'95) search an elevation map for local maxima, with three dimensional lines computed in these regions used for parametric model fits to the extracted line segments. The approach estimates the initial parameters for model fitting, but assumes that the buildings at the site can be reconstructed using a single parametric model (e.g. peaked roof model).

All these approaches assume a small class of building types (typically flat roofs), and, in the case of the ASCENDER system, where no elevation map is utilized, require several registered optical images to arrive at an accurate 3D model. This work makes no assumptions about the shape of the building rooftop, other than it is composed from a set of surface shapes defined in an existing database.

Our approach to the problem has been threefold: 1) We focused on extraction techniques that were not restricted to a small class of buildings. Instead, automatic classification of the different surface primitives are combined to reconstruct a wide class of building types. Examples are multi-level flat roofs (or single level structures with significant substructures such as air conditioner units), peaked roof buildings, juxtapositions of flat and peaked roofs, curved-roof buildings, such as Quonset huts or hangars, as well as buildings with more complex roof structures containing gables, slanted dormers or spires. 2) Our goal is to acquire 3D models even at the limits of feasibility due to noise and the size of the structures being extracted. In order to accomplish this, the use of top-down model application is used to arrive at a correct reconstruction in the presence of significant noise and ambiguities. Because of the problems inherent in the use of a single sensor type, we fuse information across sensors as necessary and make use of a database of surface types that represent geometrically feasible buildings. (51 models, in eight classes) 3) The system is fully automatic. Although user intervention can be a valuable source of information, the system attempts to segment, classify, and reconstruct the site completely automatically with as little sensitivity to parameter settings within the different algorithms as possible.

This paper presents the system, broken into the three constituent parts: bottom-up segmentation, top-down model indexing, and robust surface fitting. We demonstrate each of the algorithms through a running example, and show the results of testing the approach on two different datasets: a digital elevation map (DEM) from the ISPRS/Ascona dataset, and a subset of the Fort Hood, Texas dataset constructed via stereo-processing.

2 Segmentation

Segmentation takes place in the optical image using a perceptual grouping scheme first presented in the ASCENDER I system (Collins'95). The segmented regions from the optical image provide a focus of attention for surface reconstruction within the range image. Figure 1 shows the 450x450 pixel optical and registered range images for an area located at Fort Hood, Texas that will be used to demonstrate the reconstruction process. The range image was constructed using the UMass TERREST (Schultz'94) system from an optical pair of the region.

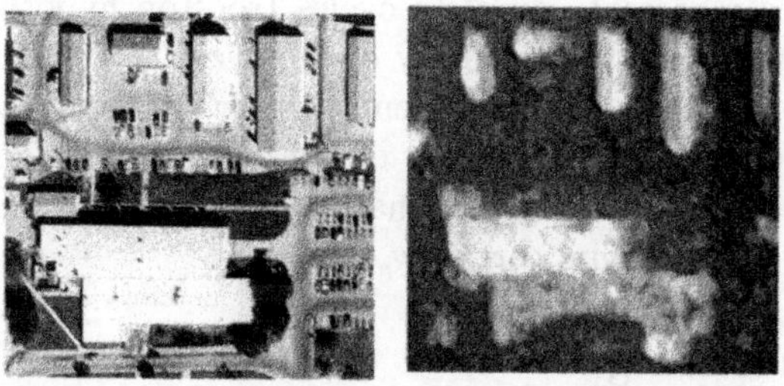

Fig. 1: An optical image and corresponding stereo range data for an area of Ft. Hood Texas. Elevation values at right are shown as image intensity.

Detection of rooftop boundaries is based on a perceptual grouping scheme (Jaynes'94) that has been shown to be effective in delineating rooftop boundaries in aerial images (Collins, Jaynes, et. al '96). The algorithm proceeds in three steps: low level feature extraction, collated feature detection via geometric constraints, and polygon extraction.

Low level features in the segmentation module are straight line segments and corners. We assume that significant rooftop surfaces can be delineated with a flat rectilinear polygons an assumption that allows the detection of the rooftop boundary of many surface shapes in a near nadir view and flat rooftops in any view. The assumption implies a search for polygons made up of straight line segments and orthogonal corners (although orthogonal corners in the world are not necessarily orthogonal in the scene when oblique views are processed). To determine a set of relevant corner hypotheses, pairs of line segments with spatially proximate endpoints are grouped together into candidate image corner features. Using the known camera pose, each potential image corner is then backprojected into a nominal Z-plane in the scene, and the hypothetical *scene corner* is tested for orthogonality.

Geometrically, collated features are sequences of geometrically grouped corners and lines that form a chain (Figure 2). Chains are a generalization of the collated features in earlier work (Huertas'80) and allow final polygons of arbitrary rectilinear shape to be constructed from the low level features.

Collated feature chains are represented by paths in a feature relation graph. Low level features (corners and line segments) are nodes in the graph, and perceptual grouping relations between these features are represented by edges in the graph. Nodes have a certainty measure that represents the confidence of the low level feature extraction routines; edges are weighted with the certainty of the grouping that the edge represents. A chain of collated features inherits an accumulated certainty measure from all the nodes and edges along its path.

High Level Polygon hypothesis extraction proceeds in two steps. First, all possible polygons are computed from the collated features. Then, polygon hypotheses are arbitrated in order to arrive at a final set of non-conflicting, high confidence rooftop polygons (Figure 2). Polygon hypotheses are simply closed chains, which can be found as cycles in the feature relation graph. All of the cycles in the feature relation graph are searched for in a depth first manner, and stored in a dependency graph where nodes represent complete cycles (rooftop hypotheses). Nodes in the dependency graph contain the certainty of the cycle that the node represents. An edge between two nodes in the dependency graph is created when cycles have low level features in common. The final set of non-overlapping rooftop polygons is the set of nodes in the dependency graph that are both independent (have no edges in common) and are of maximum certainty.

3 Surface Indexing

The segmentation module produces a set of two-dimensional closed regions within the optical data that represent high-confidence building rooftop hypotheses. To

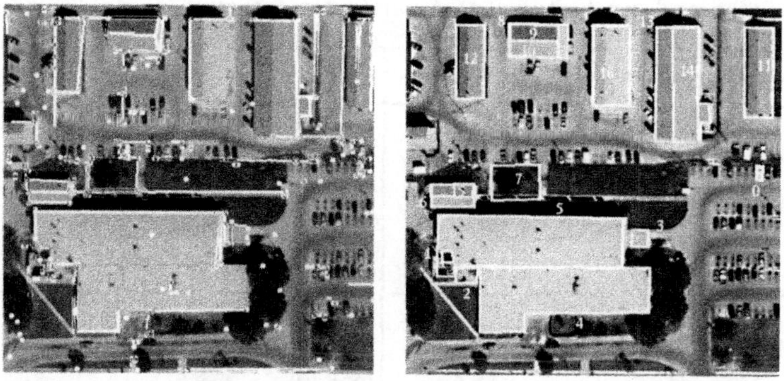

(a) (b)

Fig. 2: (a) Perceptually grouped corner and line segments that are represented in a feature relation graph. (b) The final maximally weighted set of closed cycles.

reconstruct the 3D structure, parameterized models (called surface primitives) are fit to each corresponding region in the registered range data independently. The final buildings are reconstructed as a composition of these primitives. The work here presents the case when each building is modeled as a separate primitive; however, work is underway to merge multiple models into a single complex building. The key to a reconstruction system that is able to address a wide class of building types lies in the system's ability to automatically classify the type of surface primitive associated with each region, and to constrain the parameters of a robust fit (see Section 4) sufficiently to arrive at an accurate solution. Furthermore, the method must be invariant with respect to translation, scale, and noise.

The classification scheme indexes into a database of surface primitives based on an analysis of the surface characteristics within each region. Indexing is based on the differential properties of a segmented region. This is accomplished by estimating the orientations of small surface patches, and constructing an orientation histogram that is then correlated with an existing library of roof models. These orientation histograms, sometimes called the Extended Gaussian Image (do Carmo'76), are normalized so that they are both scale and translation invariant. A detailed introduction to the Extended Gaussian Image can be found in (Horn'86).

The surface primitive database (SPD) contains a set of surface classes called *surface primitives*, such as planes, cylindrical surfaces, peaks, and spires, known to typically be part of rooftop surfaces. Associated with each surface primitive are a number of models, representing different parameterizations of each class of surface primitives. For example, the "Peak" surface primitive class is the canonical shape for a number of models in the SPD, each with a different peak angle. Corresponding orientation histograms are stored in the SPD for indexing purposes. Figure 3 shows the SPD used for the results shown here. It contains 8 different surface primitives and 51 total models. Note that the number and type of surface primitives within the SPD is unrestricted and can be tuned to a particular domain.

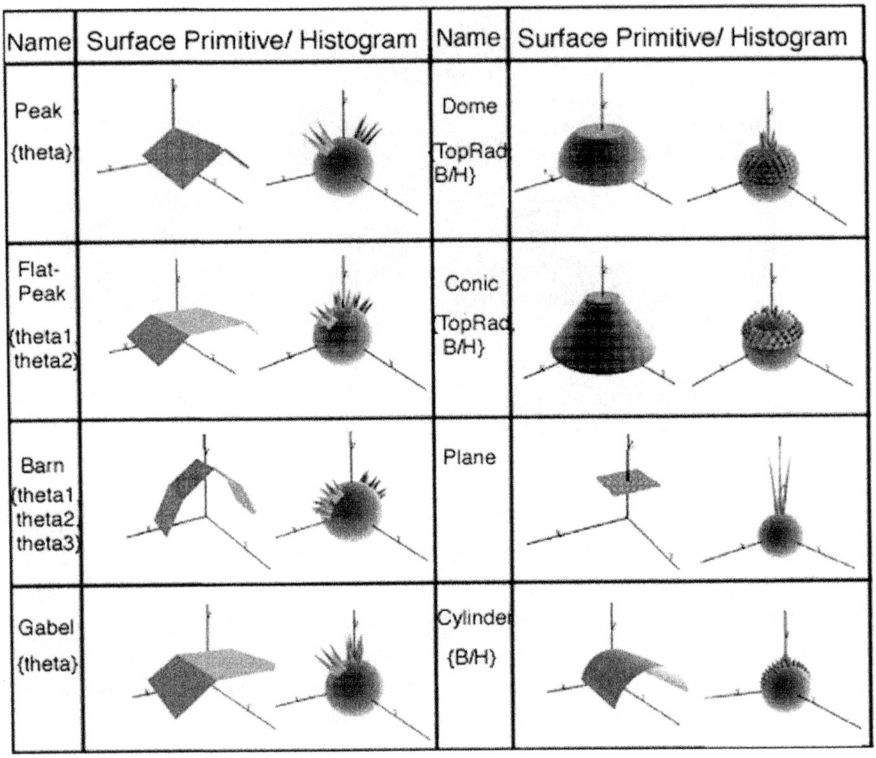

Fig. 3: The Surface Primitive Database used for indexing. Each canonical surface primitive is shown along with the orientation histogram (see text). The parameterization for each model is shown to the left. Several different models, each with unique parameters, are stored in the database. The number of different models is shown to the left.

For each of the segmented regions, an orientation histogram is constructed and correlated with the set of models stored in the SPD. Given a single region, the set of points within the region, $p_k = (i, j, z)$ are triangulated into a simple surface using the well known Delanuy algorithm [cite]. The triangulated surface is a set of triangular surface patches, $T_I = (p_1, p_2, p_3)$, where p_1, p_2, p_3 are datapoints from the original pointset. Figure 4 shows the triangulated surface that corresponds to the polygon 11 the segmentation algorithm.

The surface normal at each triangular patch is then computed as the cross product of the vectors $v_1 = \frac{(p_2 - p_1)}{||p_2 - p_1||}$ and $v_2 = \frac{p_3 - p_1}{||p_3 - p_1||}$. Because we assume the normal of the plane representing the footprint of the rooftop is aligned with the gravity vector, the surface normal pointing in the positive Z direction is used to determine the cell on the Gaussian sphere that will receive a "vote" for a particular orientation.

To avoid sensitivity problems with the method in which orientation space is discretized, votes are smoothed over the sphere via a Gaussian function. If the surface

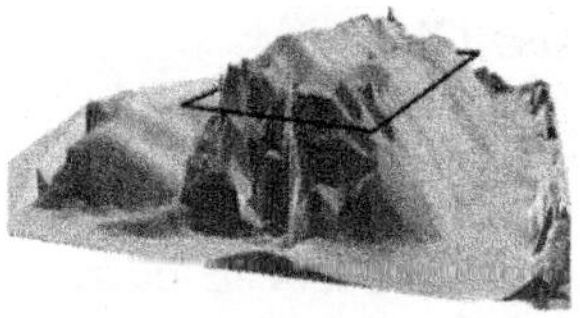

Fig. 4: The Delanuy representation of the elevation data. The polygon shown corresponds to the upper-rightmost polygon (polygon 11) from figure 2. Each surface patch "votes" in an orientation histogram for a particular orientation.

normal, N, intersects the Gaussian sphere at (x, y, z), the weighted vote is given by:

$$V(x,y,z,B) = \frac{1}{\sqrt{2\pi\sigma^2}} e^{-(D^2/\sigma^2)} \tag{1}$$

where D is the angular distance from (x, y, z) to the center of the histogram bucket, B, to receive the weighted vote. For the results shown here, the orientation histogram contains 240 buckets, reflecting a tesselation based on the semiregular icosahedron. Voting for a given vector stops when the bucket value of $V(x, y, z, B)$ falls below a threshold (0.1 for the results shown here). Figure 5 shows the histogram constructed from the range data corresponding to polygon 11 in the Fort Hood example. Histograms visualized in this way provide an interesting characterization of the noise within a surface patch. Notice, for example, that the underlying peak structure is discernible, but that number of votes for each of the "lobes" are unequal.

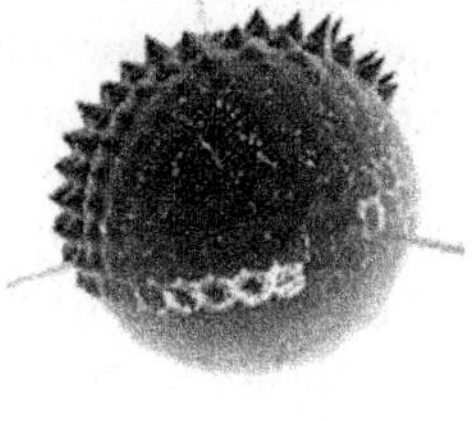

Fig. 5: The surface orientation histogram for the surface contained within the polygon shown in figure 4.

To achieve model indexing, the constructed histogram, referred to as the image histogram, is then correlated with each of the model histograms stored in the SPD. The correlation process is thus independent of scale. The normalized cross-correlation score, given by:

$$C_\theta(\mathcal{I}, \mathcal{M}) = \frac{\sum^{(\imath,\jmath)}(\mathcal{I}(\imath,\jmath) - \mu_{\mathcal{I}})(\mathcal{M}(\imath,\jmath) - \mu_{\mathcal{M}})}{(\sigma_{\mathcal{I}} * \sigma_{\mathcal{M}})} \tag{2}$$

where μ and σ represent the mean and the variance respectively.

3.1 Separability of Surface Classes

In order for the technique to be useful within the domain. The set of relevant surface classes must be separable by the correlation metric. Furthermore, the ability of the indexing technique to discriminate between classes must degrade gracefully with respect to noise. For example, figure 6 shows the downlooking optical image and the elevation map produced by a stereo-optical reconstruction process (Schultz'94). Although the peak roof structure within the row of buildings may be discernible, the clusters of trees have introduced noise that make indexing difficult. It is important that the indexing scheme is able to select the correct model under these conditions.

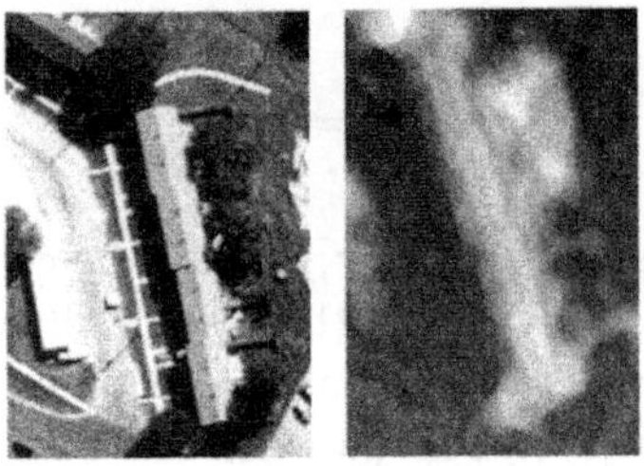

Fig. 6: Optical image and corresponding elevation data for a subregion of the Fort Hood dataset.

For each canonical model in the SPD, the set of corresponding surface primitive histograms were correlated with each other to explore the issue of intra-class separability. Figure 7 shows the correlation scores for the 35 degree peaked roof model correlated with itself and two neighboring models (25 degree and 45 degree peak angles). The 25 degree peak and 45 degree peaks reach a maximum correlation score of .51 and .60 respectivly as opposed to the maximum score of 1.0 given by the peak 35 model when perfectly aligned with itself. Clearly, then the measure is able to separate different models of the same canonical surface primitive class. However, it is important to note that the resolution of the database is limited. For the results shown here, models were separated in the database by 10 degrees. As the difference between models in the database is reduced the ability of the correlation score to separate the models is also degraded.

By correlating the same model with each of the canonical surface shapes in the database, a baseline inter-class separability of the indexing method can be measured. Figure 8 shows the results of correlating the peak 35 model with every model in the SPD. The graph depicts the model correlated with itself and the correlation response for the next two highest soring models. The flat peak at 35 degrees and the barn (35,15) model, a model with a 35 degree peak and two 15 degree slopes (see Figure 3), scored closest to the peak 35 model. As in the intra-class case, the models are separable across different surface primitives. Again, the number of different surface primtitives is limited by the ability of the correlation score to distinguish them.

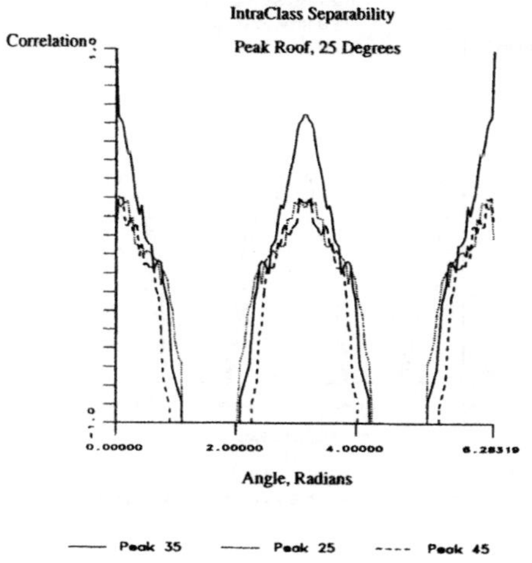

Fig. 7: A typical baseline intra-class separability using the correlation score. The Peak 35 degree model is correlated with itself and compared to the 25 degree and 45 degree peak models. (see Text)

Figure 9 shows the typical response of the correlation with the addition of uncorrelated noise. Random noise with a standard deviation of 10% of the model height, was introduced to the peak 35 model. This corrupted model was then correlated with the peak 35 model from the database, the peak 25 model, and the barn (35,15) model. As more noise is introduced the maximum correlation response for the different models slowly converges and the correlation measure is no longer able to separate the surface types. The maximum correlation response, for particular noise level was computed as the average maximum response over 100 trials.

The method not only selects the correct surface model class (peak roof, for example) based on the correlation score, but estimates the set of parameters for surface fitting (angle of the peak roof). To select the correct model orientation on the ground plane, the value of $C_\theta(\mathcal{I}, \mathcal{M})$ must be computed for many possible values of θ that represent different alignments between the spherical histograms I and M. For the results in this paper, θ was restricted to rotations about the Z-axis under the gravitational alignment assumption. The plane model, however, was allowed any possible orientation to reflect the possible presence of flat, sloped roofs or a sloping ground plane. The value of θ for which C is a maximum is stored and compared to the results of other correlations in the database. The highest scoring models are used for robust fitting during the final reconstruction phase. For the results here, the top two models are selected and fit to the data; a residual fit error is used to select the final model.

Table 1 below shows the results of correlating the histogram shown in figure 5

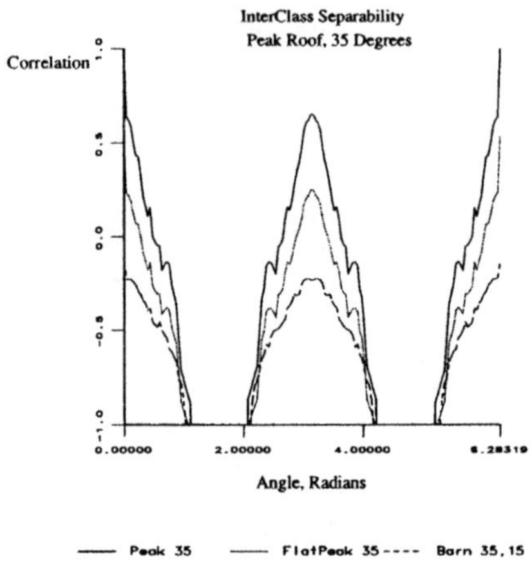

Fig. 8: A typical baseline inter-class separability using the correlation score. Peak 35 model correlated with itself and the next two top scoring models are shown. (see Text)

Model Name	C_θ
Peak (130)	0.8813 $\theta = 0.0$
Peak (150)	0.8320 $\theta = 0.0$
Flat Peak (65,65)	0.8054 $\theta = 0.0$
Flat Peak (75,75)	0.7900 $\theta = 0.0$

Polygon	Model Name	C_θ
1	Peak (130)	0.8102 $\theta = 161.0$
2	Peak (130)	0.9353 $\theta = 164.0$
3	Peak (120)	0.8322 $\theta = 162.0$
4	Peak (130)	0.8931 $\theta = 162.0$

Tab. 1: Left: Correlation response of top 4 models for polygon 11. **Right:** Model selected and correlation response for each of the hand segmented polygons (labeled from top to bottom) in figure 6.

(constructed from polygon 11) with the SPD.

The correlation score and the rotation angle of maximum response are shown for the top four models. Note that all are at $\theta = 0.0$, i.e. aligned with the Y axis. A graph depicting the correlation score of the model "Peak (130)" with the histogram produced from the polygon 11 range data is shown in figure 10. Similarly, the leftmost table 1 shows the top model, correlation score and angle of response for the classification of the four buildings in figure 6. The polygons were hand segmented in the range image in order to test the indexing scheme on perfectly segmented data.

For each of the regions produced by the segmentation module, a spherical histogram is constructed and correlated with the SPD. Figure 11 shows the model selected for each of the regions in the Fort Hood image. Once each region has been classified, a final reconstruction process determines a precise surface fit to the elevation data.

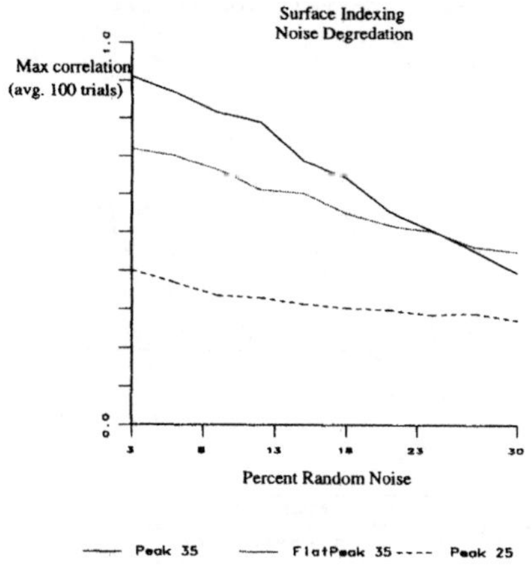

Fig. 9: Peak 35 model with random noise correlated with three different models in the SPD. As noise is added to the model, the separability of the classes diminishes. At about 25 percent random noise the flat peak 35 and peak 35 models are indistinguishable.

4 Reconstruction

Each region within the data has been indexed into the SPD to provide a set of initial models and parameters; these are then fit to the elevation data. The role of the reconstruction module is to use the initial parameters from the SPD match to determine a precise model fit to the range data.

For each polygon boundary, the set of elevation points $(x, y, z)_{uv}$ that project within the polygon are extracted from the corresponding range image. All these points are considered to be part of the initial inliers $\mathcal{I}$. The model selected from the SPD is fit to $\mathcal{I}$ using a least-median fit technique and the downhill Simplex method.

The downhill Simplex method requires a set of initial parameters to initialize the search; these are provided directly from the indexed SPD model. Each surface primitive in the SPD has an associated method for measuring the residual error from an elevation point to the surface. This information is used during the downhill simplex fit to compute an overall fit error at each step of the minimization technique. In order to avoid unusually high residual errors in the case of models with steep surfaces, residual errors are computed as the distance along the approximate surface normal from the elevation data (derived from the Delunay triangulation) to the current model estimate. The technique will converge to the model in which the median residual fit error is at a minimum.

Figure 11 shows the final reconstruction after the SPD model "Peak (130)" was fit to the data. The final peak angle (measured from plane to plane) converged to 134 degrees with a median residual error of 0.065 meters for a roof whose elevation varied by 5.9 meters.

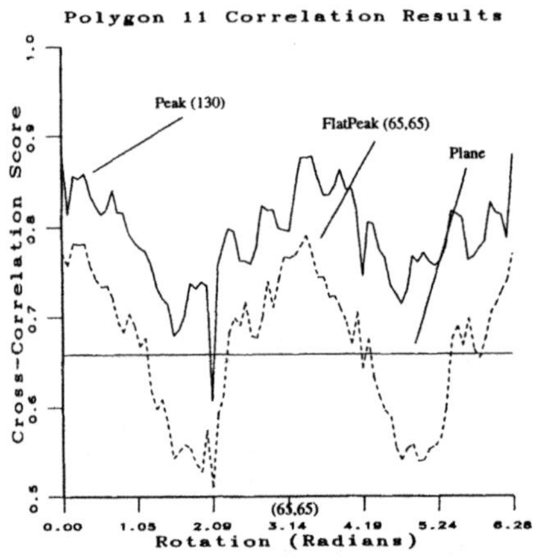

Fig. 10: The correlation response of three models (Peak, FlatPeak, and Plane, for comparison) with polygon 11. The highest scoring model other than a peaked roof primitive was the "FlatPeak (65,65)" model. The correlation score, however, clearly separates the two. The plane response is shown as the constant response since it is an orientation invariant model whose correlation score is 0.652.

Each of the regions was fit to the appropriate SPD model from the database. All remaining points in the range image were assumed to lay in a ground plane and a least-median plane was fit to determine the ground plane fit. Figure 12 shows a rendered view of the entire site model. Using the registered optical image, a texture map can be wrapped onto each of the rooftop surfaces for better visualization of the site. This was applied to the two flat roof models in Figure 12.

5 Test Results

The technique has also been applied to the Ascona/ISPRS "Flat Scene"; this scene contains several peaked roof buildings of different slopes. Several are cluttered with gabled windows and surface markings and the elevation data contains noise unavoidably introduced through the stereo reconstruction process (see Figure 14). In the faace of space restrictions, only key steps in the reconstruction process are shown.

The buildings detected are shown in figure 15. Overlapping polygons were eliminated leaving sixty four percent of the building rooftops to be passed to the indexing module for classification. Note that polygon 21 was detected but was eliminated from the reconstruction process because corresponding range data was not available. The remaining twenty regions were classified using the SPD shown in section 2. The results of the classification are shown in table 2.

The top two models selected for each region were fit to the data and the best fit was chosen for the final reconstruction. All points outside the twenty polygons are used to fit a local ground plane. Planar regions that were classified as planes and were

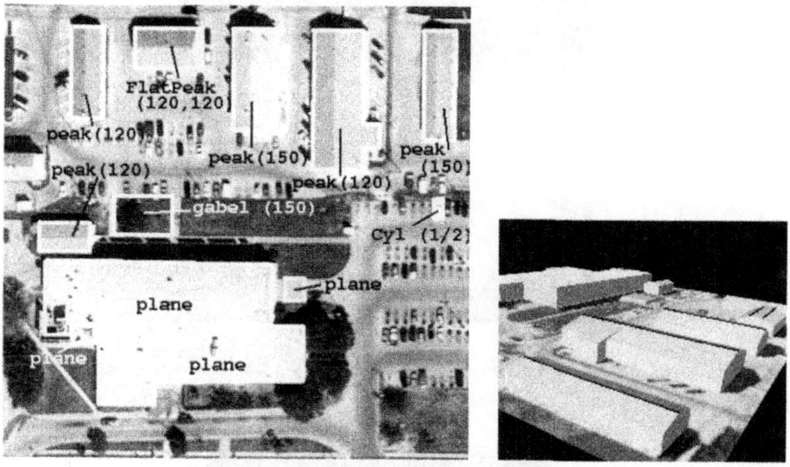

Fig. 11: Left: Classification of each region. The model with a maximum response from the library is shown. **Right:** Closeup view of reconstructed polygon 11 (building in foreground).

Polygon #	Model Name	Score	θ
1	Plane	0.9322	
2	Peak (120)	0.9104	$\theta = 121.0$
3	Plane	0.9051	
4	Plane	0.9462	
5	Plane	0.8933	
7	Plane	0.7543	$\theta = 28.0$
8	Peak (110)	0.8963	$\theta = 29.0$
9	Peak (140)	0.9502	$\theta = 118.0$
10	Plane	0.9353	
11	Peak (140)	0.9017	$\theta = 119.0$
12	Peak (120)	0.8991	$\theta = 120.0$
13	Peak (150)	0.9136	$\theta = 118.0$
14	Plane	0.8773	
15	Peak (120)	0.9114	$\theta = 119.0$
16	Peak (130)	0.9276	$\theta = 116.0$
17	Plane	0.8994	
18	Peak (130)	0.9212	$\theta = 117.0$
19	Plane	0.9102	
20	Peak (120)	0.9532	$\theta = 118.0$

Tab. 2: Classifcation results of the Ascona/ISPRS Scene.

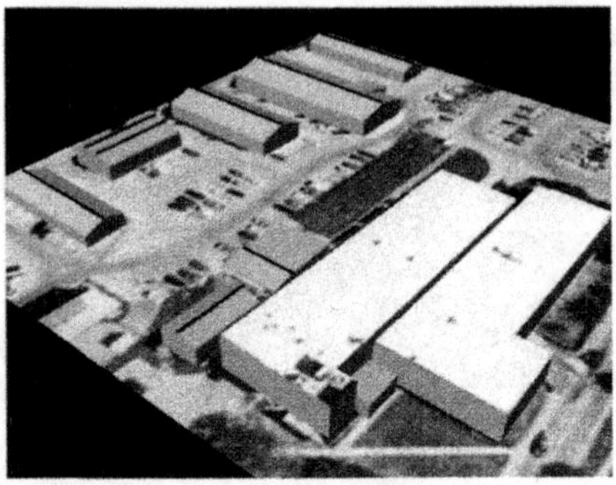

Fig. 12: Rendered view of the Fort Hood Scene. The final site supports fly through visualization.

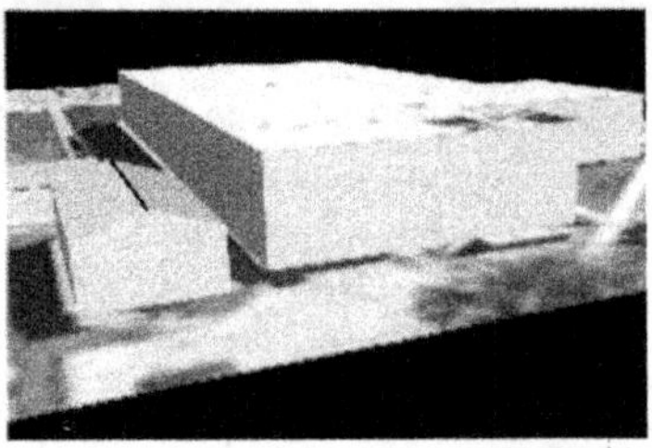

Fig. 13: Reconstructed polygons 5 (flat roof) and 6 (a peaked roof) from the Fort Hood scene.

similar to the local groundplane (for example polygon 3) were eliminated as false positives and adjusted to lay within the groundplane. Two errors not eliminated by this process are represented by polygons 5 and 7, both being sloping portions of the rooftops that are reconstructed as extruded planes. The final site model is visualized by draping the optical image over the reconstructed range image. This is shown in figure 16. A closup view of buildings 16, 18, and 20 is shown in figure 17.

References

F. Aurenhammer, "Voronoi diagrams - a survey of a fundamental geometric data structure". *ACM Computing Surveys* 1991. vol 23. pp. 345-405.

E. Baltsavias, S. Mason, and D. Stallman. "Use of DTMs/DSMs and Orthoimages to support Building Extraction", *International Workshop on Automatic Extraction of Man-Made Objects from Aerial and Space Images*, Ascona, Switzerland, April 1995.

R.Collins, Y.Cheng, C.Jaynes, F.Stolle, X.Wang, A.Hanson and E.Riseman, "Site Model Acquisition and Extension from Aerial Images," *International Conference on Computer Vision*, Cambridge, MA, June 1995, pp. 888–893.

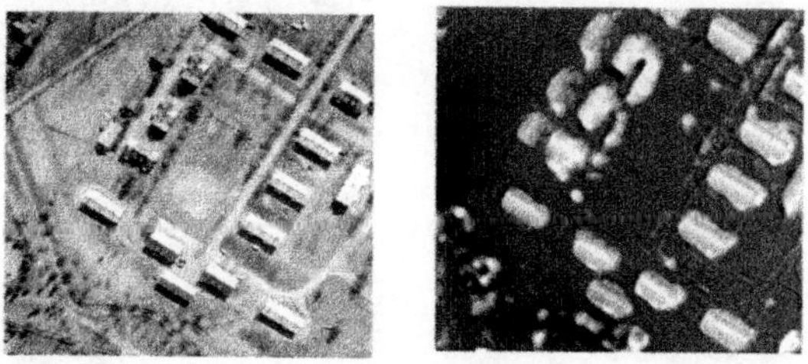

Fig. 14: Optical and range image for the Ascona/ISPRS scene.

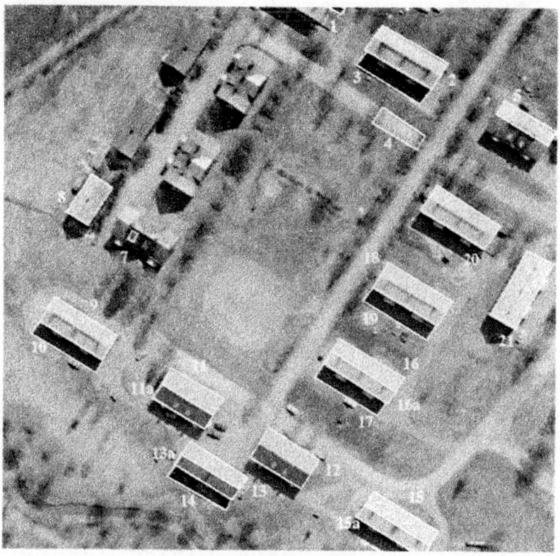

Fig. 15: Detected polygons in the Ascona scene. Polygons labeled with a subscript were eliminated due to overlap.

R. T. Collins, A. R. Hanson, E. M. Riseman, C. O. Jaynes, F. Stolle, X. Wang, Y-Q. Cheng. "UMass Progress in 3D Building Model Acquisition.", *Arpa Image Understanding Workshop,* Palm Springs, CA, Feb. 1996.

R. Collins, C. Jaynes, Y. Cheng, X. Wang, F. Stolle, H. Schultz, A. Hanson, E. Riseman. "The UMass Ascender System for 3D Site Model Construction", chapter to appear in forthcoming DARPA IU RADIUS book. Oscar Firschien (ed), 1996.

M. P. do Carmo, *Differential Geometry of Curves and Surfaces*, Prentice-Hall, Englewood Cliffs, New Jersey, 1976.

N. Haala and M. Hahn, "Data fusion for the detection and reconstruction of buildings," *International Workshop on Automatic Extraction of Man-Made Objects from Aerial and Space Images*, Ascona, Switzerland, April 1995.

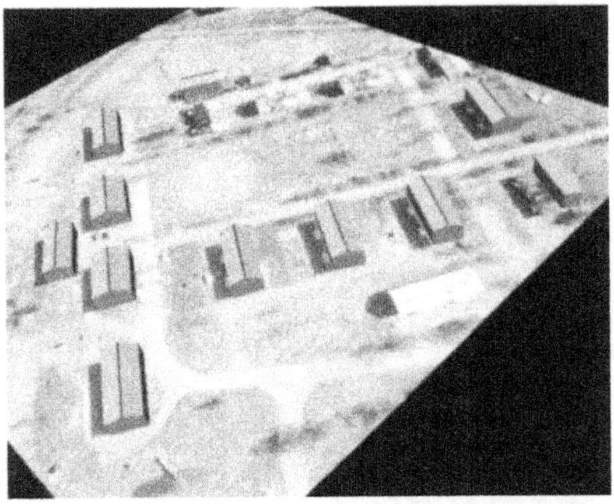

Fig. 16: The reconstructed site elevation image. The optical image is texture mapped to the new data for visualization purposes.

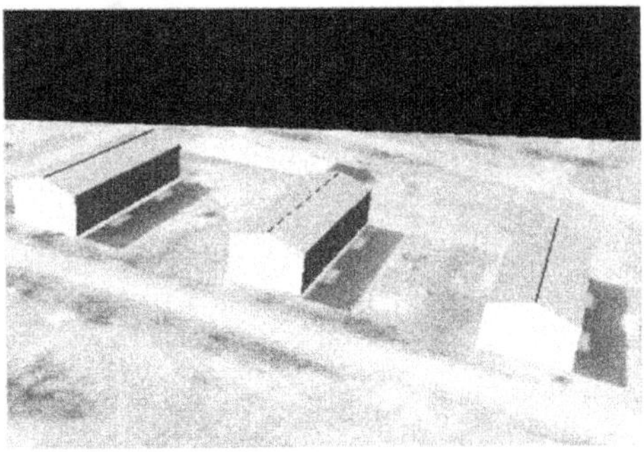

Fig. 17: Closeup view of several buildings (16, 18, and 2) reconstructed from the ISPRS/Ascona scene.

M. Herman and T. Kanade. "3D Mosaic Scene Understanding System: Incremental Reconstruction of 3D Scenes from Complex Images". *Proc. ARPA Image Understanding Workshop*, 1994.

B. Horn. *Robot Vision*, MIT Press, McGraw-Hill, Cambridge, Massachusetts, 1986.

A. Huertas and R. Nevatia. "Detecting Buildings in Aerial Images" *Computer Vision, Graphics, Image Processing.* vol. 13, 1980.

C.Jaynes, F.Stolle and R.Collins, "Task Driven Perceptual Organization for Extraction of Rooftop Polygons," *IEEE Workshop on Applications of Computer Vision,* Sarasota, FL, December 1994, pp. 152–159.

"Building Extraction and Verification from Spaceborne and Aerial Imagery using Image Understanding Fusion Techniques," *International Workshop on Automatic Extraction of Man-Made Objects from Aerial and Space Images*, Ascona, Switzerland, April 1995.

T. Matsuyama and V. Hwang, "SIGMA: A Framework for Image Understanding: Integration of Bottom-Up and Top-Do wn Processes," *Proceedings of the Ninth IJ-CAI*, Los Angeles, CA, pp. 908-915, 1985.

D. McKeown, "Toward Automatic Cartographic Feature Extraction," *Mapping and Spatial Modelling for Navigation,* Nato ASI Series, Vol. F65, p p. 149–180, 1990.

H. Schultz, "Terrain Reconstruction from Oblique Views," *Arpa Image Understanding Workshop,* Monterey, CA, Nov 1994, pp. 1001–1008.

V. Venkateswar and R. Chellapa, "Intelligent Interpretation of Aerial Images," *University of Southern California, Dept. of Electrical Engineering Technica l Report 137* March 1989.

GPSR Compliance

The European Union's (EU) General Product Safety Regulation (GPSR) is a set of rules that requires consumer products to be safe and our obligations to ensure this.

If you have any concerns about our products, you can contact us on ProductSafety@springernature.com

In case Publisher is established outside the EU, the EU authorized representative is:

Springer Nature Customer Service Center GmbH
Europaplatz 3
69115 Heidelberg, Germany

Batch number: 10732314

Printed by Printforce, the Netherlands